# Finance and Pricing 2018



Transportation Research Record: Journal of the Transportation Research Board

Volume 2672, Issue 5

ISBN: 978-1-5108-8558-5

**Printed from e-media with permission by:**

Curran Associates, Inc.
57 Morehouse Lane
Red Hook, NY 12571

**Some format issues inherent in the e-media version may also appear in this print version.**

Published by Sage Publications USA



ISBN (Print) 978-1-5108-8558-5
ISBN 2018 Printed Set (All Issues) 978-1-5108-7735-1

Printed by Curran Associates, Inc. (2019)



**Additional copies of this publication are available from:**

Curran Associates, Inc.
57 Morehouse Lane
Red Hook, NY 12571 USA
Phone: 845-758-0400
Fax: 845-758-2633
Email: curran@proceedings.com
Web: www.proceedings.com

TRR
JOURNAL OF THE TRANSPORTATION RESEARCH BOARD

Volume 2672 Number 5 December 2018

# Contents

## Finance and Pricing 2018

### Articles

# TRB
# TRANSPORTATION RESEARCH BOARD

## CORE SPONSORS

**Transportation Departments of the 50 States and the District of Columbia**

***Federal Government***

U.S. Department of Transportation
- Federal Aviation Administration (FAA)
- Federal Highway Administration
- Federal Motor Carrier Safety Administration (FMCSA)
- Federal Railroad Administration (FRA)
- Federal Transit Administration
- National Highway Traffic Safety Administration (NHTSA)
- Office of the Assistant Secretary of Transportation for Research and Technology (OST-R)

U.S. Air Force Civil Engineer Center (AFCEC)
U.S. Coast Guard (USCG)
U.S. Department of Energy
U.S. Department of the Interior (DOI)
U.S. Environmental Protection Agency (EPA)

***Nongovernmental Organizations***

American Public Transportation Association (APTA)
Association of American Railroads (AAR)
California Air Resources Board

*The National Academies of*
# SCIENCES · ENGINEERING · MEDICINE

The **National Academy of Sciences** was established in 1863 by an Act of Congress, signed by President Lincoln, as a private, nongovernmental institution to advise the nation on issues related to science and technology. Members are elected by their peers for outstanding contributions to research. Dr. Marcia McNutt is president.

The **National Academy of Engineering** was established in 1964 under the charter of the National Academy of Sciences to bring the practices of engineering to advising the nation. Members are elected by their peers for extraordinary contributions to engineering. Dr. C. D. Mote, Jr., is president.

The **National Academy of Medicine** (formerly the Institute of Medicine) was established in 1970 under the charter of the National Academy of Sciences to advise the nation on medical and health issues. Members are elected by their peers for distinguished contributions to medicine and health. Dr. Victor J. Dzau is president.

The three Academies work together as the National Academies of Sciences, Engineering, and Medicine to provide independent, objective analysis and advice to the nation and conduct other activities to solve complex problems and inform public policy decisions. The Academies also encourage education and research, recognize outstanding contributions to knowledge, and increase public understanding in matters of science, engineering, and medicine.

Learn more about the National Academies of Sciences, Engineering, and Medicine at **www.national-academies.org.**

The **Transportation Research Board** is one of seven major programs of the National Academies of Sciences, Engineering, and Medicine. The mission of the Transportation Research Board is to increase the benefits that transportation contributes to society by providing leadership in transportation innovation and progress through research and information exchange, conducted within a setting that is objective, interdisciplinary, and multimodal. The Board's varied activities annually engage about 7,000 engineers, scientists, and other transportation researchers and practitioners from the public and private sectors and academia, all of whom contribute their expertise in the public interest. The program is supported by state transportation departments, federal agencies including the component administrations of the U.S. Department of Transportation, and other organizations and individuals interested in the development of transportation.

Learn more about the Transportation Research Board at **www.TRB.org**.

TRR
JOURNAL OF THE TRANSPORTATION RESEARCH BOARD

Article

Transportation Research Record
2018, Vol. 2672(5) 1–11

DOI: 10.1177/0361198118783099
journals.sagepub.com/home/trr
SAGE

# Econometric Analysis of Monthly Peak-Hour and Total Usage Patterns of Hong Kong's Cross-Harbor Tunnels

**Chi Keung Woo[1], Kang Hua Cao[2], Yuk-shing Cheng[2], Alice Shiu[3], and Raymond Li[3]**

**Abstract**

As one of the most densely populated metropolises in the world, Hong Kong daily sees severe traffic delays at the Cross-Harbour Tunnel (CHT), though not at the Eastern Harbour Crossing (EHC) or the Western Harbour Crossing (WHC). In 2013, the Hong Kong Special Administrative Region (HKSAR) Government proposed raising the tolls of the publicly owned CHT and lowering those of the publicly owned EHC for nine vehicle types: private cars, motorcycles, taxis, three kinds of buses, and three kinds of goods vehicles. The privately owned WHC's already high tolls, however, would remain unchanged. Using monthly usage and peak-hour usage data for January 2003 through June 2015, a Generalized Leontief demand system was estimated and found that private cars, motorcycles, and goods vehicles have price-sensitive tunnel usage patterns that are also time-dependent. The usage patterns of taxis and buses, which are public transportation vehicles, are totally price-insensitive. These findings suggest that the HKSAR Government's proposed toll changes would reduce the CHT's monthly usage by 7.4%–12.2%, and peak-hour usage by 5.0–16.8%. These usage reduction estimates suggest that a time-of-use (TOU) toll design can better manage CHT congestion than the current non-TOU design.

Efficient rationing of the use of congestible roads, bridges, and tunnels is an important aspect of transportation policy (*1–7*). Like their counterparts in other Chinese metropolises (e.g., Beijing, Shanghai, Guangzhou and Shenzhen), Hong Kong drivers encounter severe traffic congestion on a daily basis, reflecting that Hong Kong is one of the most densely populated cities in the world. Encompassing an area of ~1,100 km$^2$ (1 mi$^2$ = 2.59 km$^2$), Hong Kong has a population of ~7.3 million, resulting in a high population density of ~6,636 residents per km$^2$. It has about 340 licensed vehicles for every km (1 mi = 1.62 km) of road and would become one big parking lot if most of these vehicles were on the road at the same time.

This paper is an econometric analysis of the effectiveness of the Hong Kong Special Administrative Region (HKSAR) Government's 2013 proposal to change the tolls of Hong Kong's three cross-harbor tunnels: the dual two-lane Cross-Harbour Tunnel (CHT), the dual two-lane Eastern Harbour Crossing (EHC), and the dual three-lane Western Harbour Crossing (WHC); see Figure 1. Despite an excellent public transportation system that serves about 12.5 million passenger-trips per day, Hong Kong daily sees severe traffic delays between 07:00 and 21:00 at the CHT. A cross-harbor vehicular trip via the 1.8 km CHT may take up to 30 min, thanks to long queues along the roads leading to the CHT's northern end in Kowloon Peninsula and southern end on Hong Kong Island. The same trip, however, can be completed in about 6 min (= 1.8 km distance ÷ vehicle speed of ~20 km per hour) during the uncongested hours (e.g., 02:00–06:00).

Figure 2 shows that the CHT's aggregate usage per day is about 120,000 vehicular trips in both directions, far above the design capacity of 78,000 trips under uncongested conditions. In contrast, the EHC's aggregate usage per day is below the design capacity of 78,000 trips and the WHC's is less than one third of the design

[1]Department of Asian and Policy Studies, Education University of Hong Kong, Tai Po, New Territories, Hong Kong
[2]Department of Economics, Hong Kong Baptist University, Kowloon Tong, Hong Kong
[3]School of Accounting and Finance, Hong Kong Polytechnic University, Hung Hom, Kowloon, Hong Kong

**Corresponding Author:**
Address correspondence to Kang Hua Cao: kanghuacao@hkbu.edu.hk

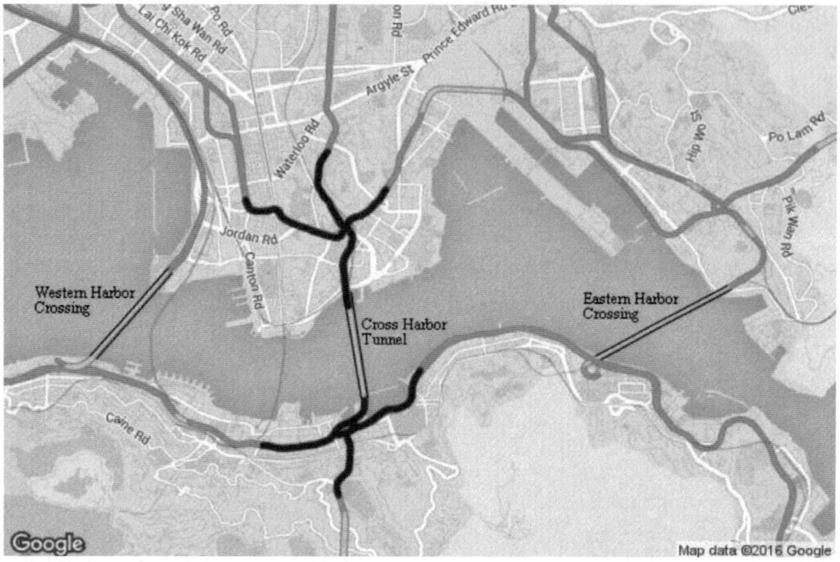


**Figure 1.** Severe congestion at Hong Kong's CHT, with the darkened line denoting the daily queues observed during rush hours along the roads leading to the CHT's northern end in Kowloon Peninsula and southern end on Hong Kong Island (*8*).

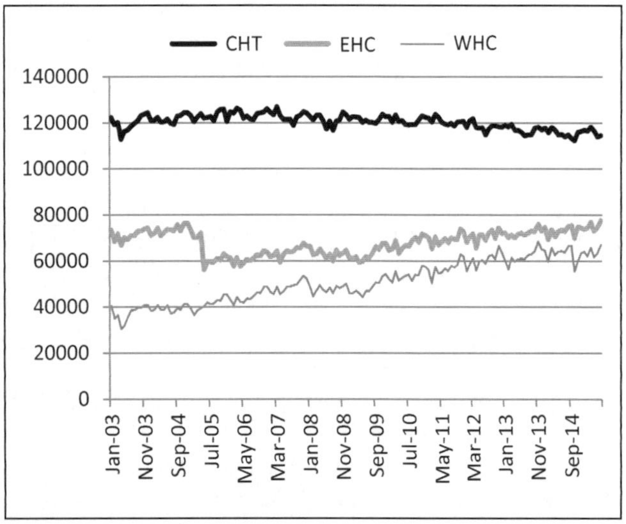


**Figure 2.** Monthly usage of the CHT, EHC, and WHC for January 2003–June 2015.

capacity of 180,000 trips. The CHT's heavy congestion is understandable, thanks to the CHT's central location as shown in Figure 1, and the low monthly average tolls shown in Figure 3.

The proposed toll changes aim to price-manage the tunnel usage patterns of nine vehicle types: private cars, taxis, motorcycles, three kinds of buses, and three kinds of goods vehicles (*9*). If implemented, the proposal would raise the vehicle-specific tolls of the publicly owned CHT and lower those of the publicly owned EHC, while keeping those of the privately owned WHC unchanged. The econometric evidence in this study is timely and relevant, underscored by the HKSAR Government's initiation in late 2015 of a 3-month public consultation on an electronic road-pricing pilot for the highly congested Central district and adjacent areas (*10*).

This paper is motivated by a presumption of price-sensitive tunnel usage patterns by vehicle type, which implies that the proposed toll changes would shift some traffic away from the heavily congested CHT to the uncongested EHC and WHC (see Figure 1). Although this presumption seems reasonable in light of the extant studies of Hong Kong's cross-harbor tunnel demands (*11–13*), a critical but unanswered question remains: can the same proposal alter each tunnel's peak-hour usage as measured by the maximum number of vehicular trips in a single hour of a given month? If the answer is "no," the proposal's effectiveness in price-managing the CHT's time-dependent congestion is very much in doubt. Therefore, the joint investigation of the three tunnels' monthly peak-hour and total usage patterns sharply differentiates this paper from (*11–13*).

The above question is answered using the monthly cross-harbor tunnel usage data for the 150-month period of January 2003 to June 2015 from the Hong Kong Transportation Department. The vehicle-specific disaggregate own- and cross-price elasticities of monthly peak-hour and total usage are estimated by tunnel and direction (northbound versus southbound). These disaggregated elasticity estimates are then used to develop each tunnel's aggregate price elasticity estimates. Finally,

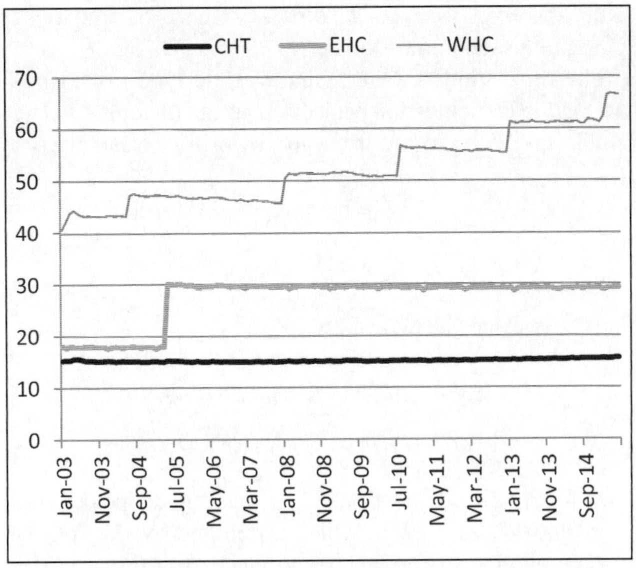


**Figure 3.** Monthly average tolls in HK$ (US$1 ≈ HK$7.8) of the CHT, EHC, and WHC for January 2003–June 2015.

the changes in each tunnel's peak-hour usage are estimated by direction, thereby addressing the CHT's time-dependent congestion problem.

This paper makes three contributions to Hong Kong's policy debates on cross-harbor tunnel congestion and the broader literature on transportation demand management. First, it presents an approach of system demand estimation that uses publicly available data to dissect a complicated real-world peak-hour congestion problem involving nine vehicle types, three tunnels, and two directions. The approach is useful when survey data collection is costly but aggregate data are readily available (*14*), yielding results that might complement empirical findings based on route-choice survey data (e.g., *12, 15–18*). Second, the paper documents that the price sensitivity of tunnel usage patterns is both time- and direction-dependent. As a result, the HKSAR Government's proposed toll changes represent a first step in price-managing the CHT's severe congestion. A future improvement is directional time-of-use (TOU) tolls (*10*). Finally, the paper provides detailed price elasticity estimates for monthly peak-hour and total usage patterns by vehicle type and direction, thereby enriching the limited evidence reported in several literature reviews (*19–23*).

## Model

The econometric formulation used here is shaped by the publicly available data. The challenge is how to best exploit these monthly data to estimate the nine vehicle types' responses to the proposed toll changes.

### *Usage Data and Variable Definitions*

The first step is to consider each tunnel's monthly total usage data by direction that corresponds to nine vehicle types, indexed here by $m = 1$ (private cars), 2 (taxis), 3 (motorcycles), 4 (light buses), 5 (single deck buses), 6 (double deck buses), 7 (goods vehicles not more than 5.5 tonnes), 8 (goods vehicles between 5.5 and 24 tonnes), and 9 (goods vehicles over 24 tonnes). Each tunnel's peak-hour usage data by direction corresponds to three size-differentiated vehicle groups, each of which has three vehicle types. The first group contains private cars, taxis, and motorcycles, the second light buses, single deck buses, and goods vehicles not more than 5.5 tonnes, and the third double deck buses, goods vehicles between 5.5 and 24 tonnes, and goods vehicles over 24 tonnes. The peak-hour usage data are derived at the vehicle type level by allocating each group's peak-hour usage among the group's three constituent vehicle types based on each vehicle type's share of the group's monthly total usage.

Let $t = 1$ for January 2003, ..., $T = 150$ for June 2015. For a given $t$ and specific direction (e.g., northbound), five usage variables for characterizing the various price elasticities are defined. For notational simplicity, the following suppresses, for the time being, the subscripts used to index a usage variable's vehicle type and direction.

The first variable is $N_{jt}$ = the vehicle type's monthly total usage of tunnel $j$. The second variable is $M_{jt}$, the vehicle type's usage in tunnel $j$'s peak hour in month $t$. The third variable is $P_{jt}$ = toll of tunnel $j$ in month $t$. The fourth and fifth variables are $Y_t$ = Hong Kong's monthly real gross domestic product (GDP) and $t$ = monthly index that captures the time trend's effect on non-toll costs.

### *Monthly Total Usage and Peak-Hour Usage of a Given Vehicle Type*

To analyze the vehicle-specific price responsiveness of $N_{jt}$ and $M_{jt}$, the process followed is the Generalized Leontief (GL) specification in (*13, 24, 25*). The GL specification is chosen because of its global properties as a flexible functional form in characterizing demands with low cross-price responsiveness. The resulting linear usage equations meaningfully link each vehicle type's $N_{jt}$ and $M_{jt}$, thereby yielding transparent calculations of aggregate peak-hour and total usage responses by tunnel and direction.

The GL equation for vehicle type's monthly total usage, $N_{jt}$, in month $t$ is

$$N_{jt} = \beta_{jj} + \sum_{k \neq j} \beta_{jk} (P_{kt}/P_{jt})^{1/2} + \psi_j Y_t + \phi_j t \quad (1)$$

and the vehicle type's peak-hour usage of tunnel $j$ is

$$M_{jt} = \alpha_{jj} + \sum\nolimits_{k \neq j} \alpha_{jk} \left(P_{kt}/P_{jt}\right)^{1/2} + \theta_j Y_t + \lambda_j t \quad (2)$$

Equations 1 and 2 state that $N_{jt}$ and $M_{jt}$ linearly depend on $(P_{kt}/P_{jt})^{1/2}$, $Y_t$, and $t$. $\beta_{jj}$ and $\alpha_{jj}$ are constants. $\{\beta_{jk}, \alpha_{jk}\}$ for $j \neq k$ are time-invariance coefficients. The effect of GDP on tunnel usage measured by $\psi_j$ and $\theta_j$ has two components: (*a*) rising income that tends to raise car ownership and drivers' trip requirements; and (*b*) rising income that tends to raise drivers' congestion costs. If (*a*) dominates (*b*), rising income tends to raise the tunnel usage.

The effect of the time trend on tunnel usage, measured by $\phi_j$ and $\lambda_j$ also has two components: (*a*) rising operations and maintenance and fuel costs that tend to increase the non-toll costs; and (*b*) an expanding transportation infrastructure (e.g., new expressways) that tends to reduce drivers' congestion costs. If (*a*) dominates (*b*), hourly tunnel usage tends to decline over time, after accounting for the effects of the monthly GDP and tunnel tolls.

### *Price Elasticities*

There are four kinds of price elasticities by direction. Because of the nonlinear nature of the elasticity formulae, sample enumeration is used via a two-step procedure to perform the elasticity calculations (*13*).

The first kind is a given vehicle type's disaggregate price elasticities for monthly total usage of tunnel *j*. Step 1 calculates

$$\eta_{jjt} = \partial \ln N_{jt}/\partial \ln P_{jt} = -1/2 \sum\nolimits_{k \neq j} \beta_{jk} \left(P_{kt}/P_{jt}\right)^{1/2}/N_{jt}$$

which is the vehicle type's own-price elasticity in month $t$. Step 2 calculates $\eta_{jj} = \Sigma_t \eta_{jjt}/T$, the equally-weighted average of $\eta_{jjt}$, to measure the vehicle-specific monthly total usage's average own-price responsiveness.

Using the same two-step procedure, the first step is to calculate

$$\eta_{jkt} = \partial \ln N_{jt}/\partial \ln P_{kt} = 1/2 \beta_{jk} \left(P_{kt}/P_{jt}\right)^{1/2}/N_{jt} \text{ for } k \neq j$$

the vehicle type's cross-price elasticity in month $t$. Then the next step calculates

$$\eta_{jk} = \sum\nolimits_t \eta_{jkt}/T$$

which is the equally-weighted average of $\eta_{jkt}$ to measure the vehicle-specific monthly total usage's average cross-price responsiveness.

The second kind is tunnel *j*'s aggregate price elasticities, each of which is denoted by $E_{jk}$ = percentage change in tunnel *j*'s aggregate usage because of a 1% change in tunnel *k*'s nine vehicle-specific tolls. Based on (*13*), $E_{jk} = (\Sigma_t \Sigma_m \eta_{jkmt} W_{jmt})/T$, where $\eta_{jkmt}$ = vehicle type *m*'s $\eta_{jkt}$ value and $W_{jmt}$ = vehicle type *m*'s share of tunnel *j*'s aggregate usage in month $t$.

The third kind is a particular vehicle type's disaggregate price elasticities for peak-hour usage of tunnel *j*. The vehicle-specific peak-hour usage's own-price elasticity is

$$\gamma_{jj} = \sum\nolimits_t \gamma_{jjt}/T, \text{where } \gamma_{jjt} = \partial \ln M_{jt}/\partial \ln P_{jt} = -1/2 \sum\nolimits_{k \neq j} \alpha_{jk} \left(P_{kt}/P_{jt}\right)^{1/2}/M_{jt}$$

The cross-price elasticities are

$$\gamma_{jk} = \sum\nolimits_t \gamma_{jkt}/T, \text{where } \gamma_{jkt} = \partial \ln M_{jt}/\partial \ln P_{kt} = 1/2\, \alpha_{jk} \left(P_{kt}/P_{jt}\right)^{1/2}/M_{jt} \text{ for } k \neq j$$

The fourth kind is tunnel *j*'s aggregate peak-hour usage elasticities, each of which is denoted by $A_{jk}$ = percentage change in tunnel *j*'s aggregate peak-hour usage because of a 1% change in tunnel *k*'s nine vehicle-specific tolls. Based on (*13*), $A_{jk} = (\Sigma_t \Sigma_m \gamma_{jkmt} S_{jmt})/T$, where $\gamma_{jkmt}$ = vehicle type *m*'s $\gamma_{jkt}$ value and $S_{jmt}$ = vehicle type *m*'s share of tunnel *j*'s aggregate peak-hour usage in month $t$.

### *Tunnel Usage Responses*

Let $P_{jm}$ denote the current toll paid by the drivers of vehicle type *m* at tunnel *j*. Further, let $P_{jm}'$ denote the proposed toll. Based on Equation 1, the effect of a proposed change in tolls on vehicle type *m*'s total usage per month of tunnel *j* for a given direction is

$$X_{jm} = \sum\nolimits_{k \neq j} \beta_{jkm} [(P_{km}'/P_{jm}')^{1/2} - (P_{km}/P_{jm})^{1/2}] \quad (3)$$

where $\beta_{jkm}$ = vehicle type *m*'s $\beta_{jk}$ coefficient for that direction. Therefore, the aggregate usage response per month by all vehicle types for tunnel *j* for a given direction is

$$X_j = \sum\nolimits_m X_{jm} \quad (4)$$

Based on Equation 2, the effect of the proposed toll change on vehicle type *m*'s peak-hour usage of tunnel *j* is

$$Z_{jm} = \sum\nolimits_{k \neq j} \alpha_{jkm} [(P_{km}'/P_{jm}')^{1/2} - (P_{km}/P_{jm})^{1/2}] \quad (5)$$

where $\alpha_{jkm}$ = vehicle type *m*'s $\alpha_{jk}$ coefficient for that direction. As there are 730 hours per month (= 8760 hours per year ÷ 12 months per year), the peak-hour response $Z_{jm}$ equals the per hour usage response ($X_{jm}$/730) when $\alpha_{jkm} = (\beta_{jkm}/730)$.

Finally, the aggregate peak-hour usage response by all vehicle types for tunnel *j* is

$$Z_j = \sum_m Z_{jm} \quad (6)$$

### *Testable Hypotheses*

To better understand congestion management via directional TOU tolls, Equations 3 and 5 are used to develop testable hypotheses based on linear restrictions on the GL system's coefficients associated with the square-rooted price ratios. Rejecting these hypotheses would suggest the new toll design's effectiveness is beyond what could be accomplished by the currently non-directional and time-invariant design.

The first hypothesis is **H1**: vehicle type $m$'s per hour usage response and peak-hour usage response are equal for a given direction. Under **H1**,

**H1A**: northbound ($\beta_{jkm}/730$) = northbound $\alpha_{jkm}$ for all $j \neq k$.

**H1B**: southbound ($\beta_{jkm}/730$) = southbound $\alpha_{jkm}$ for all $j \neq k$.

If **H1A** and **H1B** hold for *both* directions,

**H1C**: northbound ($\beta_{jkm}/730$) = northbound $\alpha_{jkm}$ *and* southbound ($\beta_{jkm}/730$) = southbound $\alpha_{jkm}$ for all $j \neq k$.

The second hypothesis is **H2**: vehicle type $m$'s directional monthly total usage responses are identical. Under **H2**, northbound $\beta_{jkm}$ = southbound $\beta_{jkm}$ for all $j \neq k$.

The third hypothesis is **H3**: vehicle type $m$'s directional peak-hour usage responses are identical. Under **H3**, northbound $\alpha_{jkm}$ = southbound $\alpha_{jkm}$ for all $j \neq k$.

The last hypothesis combines **H2** and **H3**, resulting in **H4**: vehicle type $m$'s monthly total usage *and* peak-hour usage responses do not vary by direction. Under **H4**, northbound $\beta_{jkm}$ = southbound $\beta_{jkm}$ *and* northbound $\alpha_{jkm}$ = southbound $\alpha_{jkm}$ for all $j \neq k$.

## Estimation Strategy

As there are three tunnels and nine vehicle types, each direction has 54 equations (= 27 monthly total usage equations + 27 peak-hour usage equations) to be estimated. As there are two directions, 108 is the total number of equations to be estimated. In this study, the estimation restricts $\beta_{jkm} = \beta_{kjm} \geq 0$ and $\alpha_{jkm} = \alpha_{kjm} \geq 0$ for all $j \neq k$ and $m$, thus satisfying the price concavity requirement of a well-behaved cost function.

An initial exploration reveals that it is not possible to jointly estimate the 108 usage equations because of the problem of non-convergence. As a result, each of the nine vehicle-specific GL systems is separately estimated to obtain its $\alpha_{jkm}$ and $\beta_{jkm}$ estimates. Each vehicle-specific system has 12 regressions: (*a*) three northbound monthly total usage regressions; (*b*) three southbound monthly total usage regressions; (*c*) three northbound peak-hour usage regressions; and (*d*) three southbound peak-hour usage regressions. To ascertain that the regression results are not spurious, the sample's monthly data series is first examined to see if it has a unit root and is therefore non-stationary.

Based on the Phillips-Perron (PP) test (*26*), real GDP is found to be trend-stationary and the monthly total usage and peak-hour usage are either stationary or trend-stationary at the 5% significance level used throughout the rest of this paper. The monthly square-rooted toll ratios are non-stationary. These ratios' infrequent variations, however, obviate the need to remedy the apparent non-stationarity problem because they are akin to shift dummies that move tunnel usage in response to toll changes. Therefore, the monthly data series is directly used to estimate the tunnel usage equations. The nine estimation samples by vehicle type are available from the corresponding author on request.

The next step is to apply the iterative seemingly unrelated regression (ITSUR) method in (*27*) to estimate vehicle type $m$'s monthly total usage and peak-hour usage equations differentiated by tunnel and direction:

Monthly total usage :

$$N_{jmt} = \beta_{jjm} + \sum_{k \neq j} \beta_{jkm} \left(P_{kmt}/P_{jmt}\right)^{1/2} + \psi_{jm} Y_t + \phi_{jm} t + v_{jmt} \quad (7.a)$$

Peak-hour usage :

$$M_{jmt} = \alpha_{jjm} + \sum_{k \neq j} \alpha_{jkm} \left(P_{kmt}/P_{jmt}\right)^{1/2} + \theta_{jm} Y_t + \lambda_{jm} t + \mu_{jmt} \quad (7.b)$$

The estimated $N_{jmt}$ and $M_{jmt}$ should be positive, as required by a well-behaved cost function. This requirement is met when the coefficient estimates in Equations 7.a and 7.b are all positive.

The random errors $v_{jmt}$ and $\mu_{jmt}$ on the right-hand-side (RHS) of Equations 7.a and 7.b are assumed to be contemporaneously correlated and follow an AR(3) process. The errors are contemporaneously correlated because vehicle type $m$'s usage pattern of the three tunnels is the result of the decision making by drivers of that vehicle type (e.g., private cars, motorcycles, or goods vehicles) or passengers of that vehicle type (e.g., taxis or buses). The AR(3) assumption is based on an exploration of an AR(4) process, finding that over 90% of the fourth AR parameter estimates are insignificant. These results are available on request.

## Results

### *ITSUR Regressions*

The GL systems for taxis and buses have $\alpha_{jkm}$ and $\beta_{jkm}$ ($j \neq k$) estimates that are restricted to zero. Therefore, these public transportation vehicles have price-insensitive monthly total and peak-hour usage patterns, reflecting

**Table 1.** Summary of the ITSUR/AR(3) Northbound (NB) and Southbound (SB) Regressions' Coefficient Estimates by Vehicle Type ($m$ = 1, 3, 7, 8, 9) and Tunnel ($j$ = 1, 2, 3)

| | 1. Private cars | | 3. Motorcycles | | 7. Goods vehicles not more than 5.5 tonnes | | 8. Goods vehicles between 5.5 and 24 tonnes | | 9. Goods vehicles over 24 tonnes | |
|---|---|---|---|---|---|---|---|---|---|---|
| Coefficient | NB | SB | NB | SB | NB | SB | NB | SB | NB | SB |
| Panel A. Intercepts: $\alpha_{jj}$ for peak-hour usage ($M_{jt}$) based on Equation 2 and $\beta_{jj}$ for monthly total usage ($N_{jt}$) based on Equation 1 | | | | | | | | | | |
| $\alpha_{11}$ | △ | ▽ | ▼ | ▼ | ▽ | △ | ▽ | ▲ | ▽ | ▲ |
| $\alpha_{22}$ | ▲ | ▲ | ▼ | △ | △ | ▽ | ▽ | △ | ▽ | ▽ |
| $\alpha_{33}$ | ▲ | △ | ▽ | ▽ | ▽ | △ | ▽ | ▽ | ▽ | ▼ |
| $\beta_{11}$ | △ | △ | ▼ | ▼ | ▼ | ▼ | ▼ | ▼ | ▽ | ▽ |
| $\beta_{22}$ | ▽ | ▽ | ▼ | ▼ | ▼ | ▼ | ▼ | ▼ | ▼ | ▼ |
| $\beta_{33}$ | ▼ | ▼ | ▼ | ▼ | ▼ | ▼ | ▼ | ▼ | ▼ | ▼ |
| Panel B. Square-rooted toll ratios: $\alpha_{jk}$ for peak-hour usage ($M_{jt}$) based on Equation 2 and $\beta_{jk}$ for monthly total usage ($N_{jt}$) based on Equation 1 | | | | | | | | | | |
| $\alpha_{12}$ | △ | ▲ | ▲ | ▲ | ▲ | ▲ | △ | △ | ○ | ○ |
| $\alpha_{13}$ | △ | △ | △ | ▲ | △ | △ | △ | ○ | △ | ○ |
| $\alpha_{23}$ | △ | ○ | ○ | ○ | △ | △ | ▲ | ▲ | ▲ | ▲ |
| $\beta_{12}$ | ▲ | ▲ | ▲ | ▲ | ▲ | ▲ | ▲ | ▲ | △ | △ |
| $\beta_{13}$ | ▲ | ▲ | ▲ | ▲ | ○ | ▲ | ○ | ▲ | △ | △ |
| $\beta_{23}$ | ▲ | ▲ | △ | △ | ▲ | ▲ | ▲ | ▲ | ▲ | ▲ |
| Panel C. Monthly GDP: $\theta_j$ for peak-hour usage ($M_{jt}$) based on Equation 2 and $\psi_j$ for monthly total usage ($N_{jt}$) based on Equation 1 | | | | | | | | | | |
| $\theta_1$ | △ | ▽ | ▲ | ▲ | ▼ | ▽ | ▼ | ▽ | ▽ | △ |
| $\theta_2$ | ▽ | ▽ | ▲ | ▲ | △ | △ | ▲ | ▲ | △ | ▲ |
| $\theta_3$ | △ | ▽ | ▲ | ▲ | ▲ | ▲ | ▲ | ▲ | △ | △ |
| $\psi_1$ | ▲ | ▲ | ▲ | ▲ | ▲ | ▲ | ▲ | ▲ | ▲ | ▲ |
| $\psi_2$ | ▲ | ▲ | ▲ | ▲ | ▲ | ▲ | ▲ | ▲ | ▲ | ▲ |
| $\psi_3$ | ▲ | ▲ | ▲ | ▲ | ▲ | ▲ | ▲ | ▲ | ▲ | ▲ |
| Panel D. Time trend: $\lambda_j$ for peak-hour usage ($M_{jt}$) based on Equation 2 and $\phi_j$ for monthly total usage ($N_{jt}$) based on Equation 1 | | | | | | | | | | |
| $\lambda_1$ | ▽ | ▽ | ▼ | ▼ | △ | ▽ | △ | ▽ | ▲ | △ |
| $\lambda_2$ | ▽ | ▼ | ▼ | ▼ | ▼ | ▲ | ▼ | ▼ | △ | △ |
| $\lambda_3$ | ▲ | ▲ | ▲ | ▲ | ▲ | ▲ | ▲ | ▲ | ▲ | ▲ |
| $\phi_1$ | ▼ | ▼ | ▼ | ▼ | ▼ | ▼ | ▼ | ▼ | ▽ | ▽ |
| $\phi_2$ | ▼ | ▼ | ▼ | ▼ | ▼ | ▼ | ▼ | ▼ | ▽ | ▽ |
| $\phi_3$ | ▲ | △ | △ | △ | △ | △ | △ | △ | ▲ | ▲ |

*Note*: (1) All SAS data files, programs, output listings, and logs are available from the corresponding author by email. (2) The restrictions are $\beta_{jk} = \beta_{kj} \geq 0$ and $\alpha_{jk} = \alpha_{kj} \geq 0$ for all $j \neq k$. (3) At the 5% significance level for a two-tailed $t$-test, "▲" = "positive and significant;" "▼" = "negative and significant;" "△" = "positive and insignificant;" "▽" = "negative and insignificant;" and "○" = "negative, insignificant and restricted to zero."

that their drivers do not have route choices and do not pay the tunnel tolls. This finding also suggests that passengers' route choices do not translate into empirically detectable price sensitivity of tunnel usage patterns of taxis and buses.

The rest of this subsection focuses on the results of the 30 monthly total usage and 30 peak-hour usage regressions for private cars, motorcycles, and goods vehicles. Table 1 summarizes the voluminous coefficient estimates. Panel A indicates that some intercept estimates are significantly negative. Thus, the estimated regressions are used to calculate the within-sample predictions of the peak-hour and monthly total usage. These predictions are strictly positive for all observations, as required by a well-behaved cost function.

Panel B reports the estimates for $\alpha_{jkm}$ and $\beta_{jkm}$ ($j \neq k$). Ten of the 30 $\alpha_{jkm}$ estimates are significantly positive for the peak-hour usage regressions. Twenty-two of the 30 $\beta_{jkm}$ estimates are significantly positive for the monthly total usage regressions. Although there are negative $\alpha_{jkm}$ and $\beta_{jkm}$ coefficient estimates, they are all insignificant and therefore restricted to zero. Therefore, Panel B suggests the empirical plausibility of the peak-hour and monthly total usage regressions for private cars, motorcycles, and goods vehicles.

The significant coefficient estimates for monthly GDP in Panel C suggest that rising GDP tends to raise monthly total usage, though less so for peak-hour usage. The time trend's coefficient estimates in Panel D have mixed significance levels and signs. Therefore, they do not tell a uniform story of how the monthly total usage and peak-hour usage may move over time, after accounting for the effects of the tunnel tolls and monthly GDP.

### *Hypothesis Testing*

The second to fourth columns of Table 2 report the $p$-values of the Wald statistics for testing **H1A** to **H1C**,

**Table 2.** Summary of the $p$-Values of the Wald Statistics for Testing H1A to H4

| Vehicle type | H1A: Northbound $(\beta_{jk}/730)$ = Northbound $\alpha_{jk}$ for all $j \neq k$ | H1B: Southbound $(\beta_{jk}/730)$ = Southbound $\alpha_{jk}$ for all $j \neq k$ | H1C: H1A plus H1B | H2: Northbound $\beta_{jk}$ = Southbound $\beta_{jk}$ for all $j \neq k$ | H3: Northbound $\alpha_{jk}$ = Southbound $\alpha_{jk}$ for all $j \neq k$ | H4: H2 plus H3 |
|---|---|---|---|---|---|---|
| 1. Private cars | 0.7343 | <0.0001 | <0.0001 | 0.0006 | 0.6377 | 0.0045 |
| 3. Motorcycles | 0.0015 | 0.0514 | 0.0057 | 0.0963 | Unavailable | Unavailable |
| 7. Goods vehicles not more than 5.5 tonnes | 0.0933 | 0.2084 | 0.0972 | 0.0048 | 0.3416 | 0.0075 |
| 8. Goods vehicles between 5.5 and 24 tonnes | 0.0757 | 0.0340 | 0.0144 | 0.0289 | 0.4025 | 0.0623 |
| 9. Goods vehicles over 24 tonnes | 0.1102 | 0.0267 | 0.0503 | 0.9412 | 0.4953 | 0.7761 |

*Note:* PROC MODEL of SAS (2010) does not provide the $p$-values for **H3** and **H4** when motorcycles' northbound and southbound $\alpha_{23}$ estimates are restricted to zero in the ITSUR/AR(3) estimation. $p$-Values $\leq .05$ in bold.

**Table 3.** Disaggregate Elasticity ($\eta_{jk}$) Estimates Based on Significant $\beta_{jk}$ Estimates for Monthly Total Usage ($N_{jt}$) by Direction and Vehicle Type

| Vehicle type | $\eta_{11}$ | $\eta_{12}$ | $\eta_{13}$ | $\eta_{21}$ | $\eta_{22}$ | $\eta_{23}$ | $\eta_{31}$ | $\eta_{32}$ | $\eta_{33}$ |
|---|---|---|---|---|---|---|---|---|---|
| Panel A: Northbound | | | | | | | | | |
| 1. Private cars | −0.31 | 0.08 | 0.23 | 0.08 | −0.32 | 0.24 | 0.17 | 0.18 | −0.35 |
| 3. Motorcycles | −0.67 | 0.41 | 0.26 | 0.60 | −0.60 | 0.00 | 0.88 | 0.00 | −0.88 |
| 7. Goods vehicles not more than 5.5 tonnes | −0.32 | 0.32 | 0.00 | 0.32 | −0.49 | 0.17 | 0.00 | 0.21 | −0.21 |
| 8. Goods vehicles between 5.5 and 24 tonnes | −0.30 | 0.30 | 0.00 | 0.22 | −0.41 | 0.19 | 0.00 | 0.28 | −0.28 |
| 9. Goods vehicles over 24 tonnes | 0.00 | 0.00 | 0.00 | 0.00 | −0.53 | 0.53 | 0.00 | 0.88 | −0.88 |
| Panel B: Southbound | | | | | | | | | |
| 1. Private cars | −0.35 | 0.10 | 0.24 | 0.11 | −0.32 | 0.20 | 0.20 | 0.15 | −0.35 |
| 3. Motorcycles | −0.63 | 0.39 | 0.24 | 0.60 | −0.60 | 0.00 | 0.91 | 0.00 | −0.91 |
| 7. Goods vehicles not more than 5.5 tonnes | −0.35 | 0.31 | 0.04 | 0.32 | −0.47 | 0.15 | 0.05 | 0.18 | −0.23 |
| 8. Goods vehicles between 5.5 and 24 tonnes | −0.32 | 0.27 | 0.05 | 0.21 | −0.40 | 0.19 | 0.05 | 0.24 | −0.29 |
| 9. Goods vehicles over 24 tonnes | 0.00 | 0.00 | 0.00 | 0.00 | −0.52 | 0.52 | 0.00 | 0.67 | −0.67 |

*Note:* The elasticity estimates for taxis and buses are all equal to zero because their $\beta_{jk}$ estimates ($j \neq k$) have been restricted to zero. Further, the insignificant $\beta_{jk}$ estimates for the vehicle types in this table are conservatively set to zero.

suggesting that the peak-hour usage and per hour usage of private cars, motorcycles, and goods vehicles have different price sensitivities. The $p$-values for testing **H2** indicate that the price sensitivity of private cars' and goods vehicles' monthly total usage varies by direction. The $p$-values for testing **H3**, however, suggest that the peak-hour usage's price sensitivity is largely non-directional. The $p$-values for testing **H4** indicate directionally differentiated responses by private cars and goods vehicles. When taken together, these findings suggest the potential usefulness of directional TOU tolls to ease the CHT's congestion.

## *Price Elasticity Estimates*

When calculating the price elasticities reported here and the tunnel usage responses in the next section, the insignificant $\alpha_{jkm}$ and $\beta_{jkm}$ ($j \neq k$) estimates were conservatively set to zero for two reasons. First, using the unadjusted $\alpha_{jkm}$ estimates leads to implausibly large peak-hour elasticity estimates. Second, the response estimates given by Equations 4 and 6 are insignificant when the $\alpha_{jkm}$ and $\beta_{jkm}$ ($j \neq k$) estimates are insignificant. Setting the insignificant $\alpha_{jkm}$ and $\beta_{jkm}$ estimates to zero pre-empts allegations of an imprudent use of large but insignificant response estimates to substantiate price-management's effectiveness in easing CHT congestion.

Table 3 reports the own- and cross-price elasticity estimates for the monthly total usage of private cars, motorcycles, and goods vehicles, showing that these vehicle types have price-inelastic total usage patterns. Table 4 reports the own- and cross-price elasticity estimates for the peak-hour usage patterns of private cars, motorcycles, and goods vehicles. These peak-hour elasticity estimates are generally smaller in size than those in

**Table 4.** Disaggregate Elasticity ($\gamma_{jk}$) Estimates Based on Significant $\alpha_{jk}$ Estimates for Peak-Hour Usage ($M_{jt}$) by Direction and Vehicle Type

| Vehicle type | $\gamma_{11}$ | $\gamma_{12}$ | $\gamma_{13}$ | $\gamma_{21}$ | $\gamma_{22}$ | $\gamma_{23}$ | $\gamma_{31}$ | $\gamma_{32}$ | $\gamma_{33}$ |
|---|---|---|---|---|---|---|---|---|---|
| Panel A: Northbound | | | | | | | | | |
| 1. Private cars | 0.00 | 0.00 | 0.00 | 0.00 | 0.00 | 0.00 | 0.00 | 0.00 | 0.00 |
| 3. Motorcycles | −0.41 | 0.41 | 0.00 | 0.43 | −0.43 | 0.00 | 0.00 | 0.00 | 0.00 |
| 7. Goods vehicles not more than 5.5 tonnes | −1.06 | 1.06 | 0.00 | 0.24 | −0.24 | 0.00 | 0.00 | 0.00 | 0.00 |
| 8. Goods vehicles between 5.5 and 24 tonnes | 0.00 | 0.00 | 0.00 | 0.00 | −0.23 | 0.23 | 0.00 | 0.32 | −0.32 |
| 9. Goods vehicles over 24 tonnes | 0.00 | 0.00 | 0.00 | 0.00 | −0.56 | 0.56 | 0.00 | 0.91 | −0.91 |
| Panel B: Southbound | | | | | | | | | |
| 1. Private cars | −0.21 | 0.21 | 0.00 | 0.13 | −0.13 | 0.00 | 0.00 | 0.00 | 0.00 |
| 3. Motorcycles | −0.78 | 0.43 | 0.34 | 0.39 | −0.39 | 0.00 | 0.71 | 0.00 | −0.71 |
| 7. Goods vehicles not more than 5.5 tonnes | −1.16 | 1.16 | 0.00 | 0.37 | −0.37 | 0.00 | 0.00 | 0.00 | 0.00 |
| 8. Goods vehicles between 5.5 and 24 tonnes | 0.00 | 0.00 | 0.00 | 0.00 | −0.20 | 0.20 | 0.00 | 0.24 | −0.24 |
| 9. Goods vehicles over 24 tonnes | 0.00 | 0.00 | 0.00 | 0.00 | −0.52 | 0.52 | 0.00 | 0.63 | −0.63 |

*Note*: The elasticity estimates for taxis and buses are all equal to zero because their $\alpha_{jk}$ estimates ($j \neq k$) have been restricted to zero. Further, the insignificant $\alpha_{jk}$ estimates for the vehicle types in this table are conservatively set to zero.

**Table 5.** Aggregate Elasticity ($E_{jk}$) Estimates for Monthly Total Usage and Peak-Hour Usage

| Tunnel ID | 1. Cross-Harbour Tunnel (CHT) | 2. Eastern Harbour Crossing (EHC) | 3. Western Harbour Crossing (WHC) |
|---|---|---|---|
| Panel A: Total northbound usage | | | |
| 1. CHT | −0.2065 | 0.1113 | 0.0951 |
| 2. EHC | 0.1120 | −0.2749 | 0.1629 |
| 3. WHC | 0.1002 | 0.1179 | −0.2182 |
| Panel B: Total southbound usage | | | |
| 1. CHT | −0.2276 | 0.1183 | 0.1093 |
| 2. EHC | 0.1313 | −0.2716 | 0.1403 |
| 3. WHC | 0.1152 | 0.1002 | −0.2154 |
| Panel C: Peak-hour northbound usage | | | |
| 1. CHT | −0.1056 | 0.1056 | 0.0000 |
| 2. EHC | 0.0503 | −0.0595 | 0.0092 |
| 3. WHC | 0.0000 | 0.0060 | −0.0057 |
| Panel D: Peak-hour southbound usage | | | |
| 1. CHT | −0.2342 | 0.2189 | 0.0153 |
| 2. EHC | 0.1313 | −0.1422 | 0.0109 |
| 3. WHC | 0.0063 | 0.0070 | −0.0137 |

Table 3. This makes sense because the peak-hour estimates correspond to the maximal tunnel demands of the drivers of these vehicle types.

Panels A and B of Table 5 report the aggregate own- and cross-price elasticity estimates for each tunnel's total usage, showing that all three tunnels exhibit price-inelastic usage patterns. Panels C and D report the aggregate own- and cross-price elasticity estimates for each tunnel's peak-hour usage. These peak-hour elasticity estimates are generally smaller in size than those in Panels A and B.

## *Aggregate Responses to the Proposed Toll Changes*

Panel A of Table 6 presents the HKSAR Government's three options for the CHT's and EHC's toll changes (*9*). Some of the relative toll changes are large, as evidenced by Option A's 25% (= HK\$5/HK\$20) increase for the CHT's toll for private cars and 87% (= HK\$26/HK\$30) increase for the CHT's toll for goods vehicles over 24 tonnes.

Based on Equation 4, Panel B of Table 6 summarizes each option's estimated effects on each tunnel's monthly total usage by direction. It shows that Option A is estimated to reduce the CHT's total usage by between 11.3% and 12.0% and increase the EHC's by between 14.9% and 15.1%. Its estimated effect on the WHC is negligible. The estimated effects of the other options are smaller than, but qualitatively similar to, those of Option A.

Based on Equation 6, Panel C of Table 6 summarizes each option's estimated effects on each tunnel's monthly

**Table 6.** Aggregate Responses to the Toll Changes Proposed by the HKSAR Government

| | Option A | | | Option B | | | Option C | | |
|---|---|---|---|---|---|---|---|---|---|
| Vehicle type/direction | CHT | EHC | WHC | CHT | EHC | WHC | CHT | EHC | WHC |
| Panel A: Proposed tolls (HK$/vehicle trip) in Hong Kong Transport and Housing Bureau (2013, p.13) (9), with changes from the January 2016 tolls in ( ) | | | | | | | | | |
| 1. Private cars | 25 ( + 5) | 20 (−5) | 60 (0) | 25 ( + 5) | 20 (−5) | 60 (0) | 30 ( + 10) | 20(−5) | 60 (0) |
| 2. Taxis | 19 ( + 9) | 15 (−10) | 55 (0) | 13 ( + 3) | 20 (−5) | 55 (0) | 10 (0) | 15(−10) | 55 (0) |
| 3. Motorcycles | 12 ( + 4) | 9 (−4) | 25 (0) | 10 ( + 2) | 10 (−3) | 25 (0) | 12 ( + 4) | 9(−4) | 25 (0) |
| 4. Light buses | 25 ( + 15) | 20 (−18) | 70 (0) | 13 ( + 3) | 30 (−8) | 70 (0) | 10 (0) | 38 (0) | 70 (0) |
| 5. Single deck buses | 31 ( + 21) | 25 (−25) | 110 (0) | 13 ( + 3) | 40 (−10) | 110 (0) | 10 (0) | 50 (0) | 110 (0) |
| 6. Double deck buses | 47 ( + 32) | 38 (−37) | 155 (0) | 19 ( + 4) | 60 (−15) | 155 (0) | 15 (0) | 75 (0) | 155 (0) |
| 7. Goods vehicles not more than 5.5 tonnes | 28 ( + 13) | 23 (−15) | 70 (0) | 19 ( + 4) | 30 (−8) | 70 (0) | 19 ( + 4) | 23 (−15) | 70 (0) |
| 8. Goods vehicles between 5.5 and 24 tonnes | 38 ( + 18) | 30 (−20) | 95 (0) | 25 ( + 5) | 40 (−10) | 95 (0) | 25 ( + 5) | 30 (−20) | 95 (0) |
| 9. Goods vehicles over 24 tonnes | 56 ( + 26) | 45 (−30) | 125 (0) | 38 ( + 8) | 60 (−15) | 125 (0) | 38 ( + 8) | 45 (−30) | 125 (0) |
| Panel B: Change in average monthly total usage (= Annual Sum of Monthly Total Usage ÷ 12 Months) of all vehicle types; percentage change from the July 2014–June 2015 value in ( ) | | | | | | | | | |
| Northbound | −195746 | 171227 | −4855 | −129164 | 102324 | −2015 | −199055 | 149623 | 7643 |
| | (−11.3%) | (14.9%) | (−0.5%) | (−7.4%) | (8.9%) | (−0.2%) | (−11.5%) | (13.0%) | (0.8%) |
| Southbound | −211428 | 167448 | 2268 | −142047 | 102169 | 2878 | −215854 | 151277 | 14022 |
| | (−12.0%) | (15.1%) | (0.2%) | (−8.1%) | (9.2%) | (0.3%) | (−12.2%) | (13.6%) | (1.5%) |
| Panel C: Change in average peak-hour usage (= Annual Sum of Peak-Hour Usage ÷ 12 Months) of all vehicle types; percentage change from the July 2014–June 2015 value in ( ) | | | | | | | | | |
| Northbound | −290 | 228 | −7 | −150 | 90 | −3 | −220 | 158 | −7 |
| | (−9.6%) | (7.2%) | (−0.2%) | (−5.0%) | (2.9%) | (−0.1%) | (−7.3%) | (5.0%) | (−0.2%) |
| Southbound | −521 | 414 | −2 | −322 | 229 | 0 | −462 | 369 | −2 |
| | (−16.8%) | (12.9%) | (−0.1%) | (−10.3%) | (7.1%) | (−0.0%) | (−14.9%) | (11.5%) | (−0.1%) |

peak-hour usage by direction. It shows that Option A is estimated to reduce the CHT's peak-hour usage by about 9.6–16.8% and increase the EHC's by 7.2–12.9%. Its estimated effect on the WHC is negligible. The other options' estimated effects are smaller than Option A's.

## Conclusion

Using the January 2003–June 2015 monthly peak-hour and total usage data for nine vehicle types, it is possible to document that Hong Kong can price-manage its CHT's congestion because the three cross-harbor tunnels' aggregate usage patterns are found to have discernible price responsiveness. The three tunnels' disaggregate price elasticity estimates, however, suggest price-inelastic monthly total usage patterns by vehicle type. The most price-sensitive total usage pattern belongs to motorcycles, followed by private cars and goods vehicles. Taxis and buses, which are public transportation vehicles, do not have price-sensitive monthly total usage patterns. Further, the aggregate elasticity estimates indicate that the three tunnels' monthly total usage patterns are price-inelastic, implying that the proposed toll changes can only modestly shift Hong Kong's cross-harbor tunnel traffic. Finally, the monthly peak-hour usage patterns' disaggregate and aggregate price elasticity estimates are generally smaller in size than those of the monthly total usage patterns. Therefore, the effectiveness of price-managing the CHT's peak-hour congestion may be improved by implementing directional TOU tolls as part of the HKSAR Government's pilot of electronic road pricing.

### Author Contributions

The authors confirm contribution to the paper as follows: study conception and design: CK Woo, KH Cao, YS Cheng; data collection: YS Cheng, KH Cao; analysis and interpretation of results: CK Woo, KH Cao, R Li; draft manuscript preparation: CK Woo, A Shiu. All authors reviewed the results and approved the final version of the manuscript.

### References

1. Vickrey, W. S. Optimization of Traffic and Facilities. *Journal of Transport Economics and Policy*, Vol. 1, No. 2, 1967, pp. 123–136.
2. Vickrey, W. S. Congestion Theory and Transport Investment. *American Economic Review*, Vol. 59, No. 2, 1969, pp. 251–260.
3. Pretty, R. L. Road Pricing: A Solution for Hong Kong? *Transportation Research Part A: General*, Vol. 22, No. 5, 1988, pp. 319–327.
4. Meyer, M. D. Demand Management as an Element of Transportation Policy: Using Carrots and Sticks to Influence Travel Behavior. *Transportation Research Part A: Policy and Practice*, Vol. 33, No. 7–8, 1999, pp. 575–599.
5. May, A. D., and D. S. Milne. Effects of Alternative Road Pricing Systems on Network Performance. *Transportation Research Part A: Policy and Practice*, Vol. 34, No. 6, 2000, pp. 407–436.
6. Fosgerau, M., and A. de Palma. The Dynamics of Urban Traffic Congestion and the Price of Parking. *Journal of Public Economics*, Vol. 105, 2013, pp. 106–115.
7. Green, C. P., J. S. Heywood, and M. Navarro. Traffic Accidents and the London Congestion Charge. *Journal of Public Economics*, Vol. 133, 2016, pp. 11–22.
8. Wilbur Smith Associates Limited. *Consultancy Services for Providing Expert Advice on Rationalising the Utilization of Road Harbour Crossings*. Hong Kong Transport and Housing Bureau, 2010. http://www.thb.gov.hk/eng/policy/transport/policy/consultation/RHC_Final_Report_Sep2010.pdf. Accessed June 25, 2013.
9. Hong Kong Transport and Housing Bureau. *Proposed Measures to Improve the Traffic Distribution Among the Road Harbour Crossings: Public Consultation*. The Government of the Hong Kong Special Administrative Region, 2013.
10. Hong Kong Transport Department. *Electronic Road Pricing Pilot Scheme in Central and Its Adjacent Areas (Public Engagement Document)*. The Government of the Hong Kong Special Administrative Region, 2015.
11. Loo, B. P. Y. Tunnel Traffic and Toll Elasticities in Hong Kong: Some Recent Evidence for International Comparisons. *Environment and Planning A*, Vol. 35, No. 2, 2003, pp. 249–276.
12. Hau, T. D., B. P. Y. Loo, K. I. Wong, and S. C. Wong. An Estimation of Efficient Time-Varying Tolls for Cross Harbor Tunnels in Hong Kong. *The Singapore Economic Review*, Vol. 56, 2011, pp. 467–488.
13. Woo, C. K., Y. S. Cheng, R. Li, A. Shiu, S. T. Ho, and I. Horowitz. Can Hong Kong Price-Manage Its Cross-Harbor-Tunnel Congestion? *Transportation Research Part A: Policy and Practice*, Vol. 82, 2015, pp. 94–109.
14. Nam, K. C. A Study on the Estimation and Aggregation of Disaggregate Models of Mode Choice for Freight Transport. *Transportation Research Part E: Logistics and Transportation Review*, Vol. 33, No. 3, 1997, pp. 223–231.
15. Burris, M. W., and R. M. Pendyala. Discrete Choice Models of Traveler Participation in Differential Time of Day Pricing Programs. *Transport Policy*, Vol. 9, No. 3, 2002, pp. 241–251.
16. Olszewski, P., and L. Xie. Modelling the Effects of Road Pricing on Traffic in Singapore. *Transportation Research Part A: Policy and Practice*, Vol. 39. No. 7–9, 2005, pp. 755–772.
17. Washbrook, K., W. Haider, and M. Jaccard. Estimating Commuter Mode Choice: A Discrete Choice Analysis of the Impact of Road Pricing and Parking Charges. *Transportation*, Vol. 33, No. 6, 2006, pp. 621–639.
18. Train, K., and W. W. Wilson. Estimation on Stated-Preference Experiments Constructed from Revealed-Preference

Choices. *Transportation Research Part B: Methodological*, Vol. 42, No. 3, 2008, pp. 191–203.
19. Goodwin, P. B. A Review of New Demand Elasticities with Special Reference to Short and Long Run Effects of Price Changes. *Journal of Transport Economics and Policy*, Vol. 26, No. 2, 1992, pp. 155–169.
20. Oum, T. H., W. G. Waters, and J. S. Yong. Concepts of Price Elasticities of Transport Demand and Recent Empirical Estimates. *Journal of Transport Economics and Policy*, Vol. 26, No. 2, 1992, pp. 139–154.
21. Graham, D. J., and S. Glaister. Road Traffic Demand Elasticity Estimates: A Review. *Transport Reviews*, Vol. 24, No. 3, 2004, pp. 261–274.
22. Litman, T. Transit Price Elasticities and Cross-Elasticities. *Journal of Public Transportation*, Vol. 7, No. 2, 2004, pp. 37–58.
23. Litman, T. *Understanding Transport Demands and Elasticities: How Prices and Other Factors Affect Travel Behavior*. Victoria Transport Policy Institute, 2013. http://www.vtpi.org/elasticities.pdf. Accessed June 20, 2013.
24. Diewert, W. E. An Application of the Shephard Duality Theorem: A Generalized Leontief Production Function. *Journal of Political Economy*, Vol. 79, No. 3, 1971, pp. 481–507.
25. Caves, D. W., and L. R. Christensen. Global Properties of Flexible Functional Forms. *American Economic Review*, Vol. 70, No. 3, 1980, pp. 422–432.
26. Phillips, P. C. B., and P. Perron. Testing for a Unit Root in Time Series Regression. *Biometrika*, Vol. 75, No. 2, 1988, pp. 335–346.
27. SAS. *SAS/ETS 9.22 User's Guide*. SAS Institute Inc, Cary, NC, 2010.

*The Standing Committee on Congestion Pricing (ABE25) peer-reviewed this paper (18-00363).*

Article

TRR
JOURNAL OF THE TRANSPORTATION RESEARCH BOARD

Transportation Research Record
2018, Vol. 2672(5) 12–22

DOI: 10.1177/0361198118786637
journals.sagepub.com/home/trr
SAGE

# Choice of High Occupancy/Toll Lanes under Alternative Pricing Strategies

**Michael Janson[1] and David Levinson[2]**

**Abstract**

High occupancy/toll (HOT) lanes typically vary tolls charged to single occupant vehicles, with the toll increasing during congested periods. The toll is usually tied to time of day or to the density of vehicles in the HOT lane. The purpose of raising the toll with congestion is to discourage demand sufficiently to maintain travel speeds in the HOT lane. However, it has been demonstrated that the HOT toll may act as a signal of downstream congestion (in both general purpose (GP) and HOT lanes), causing an increase in demand for the HOT lane, at least at lower prices. This paper develops a model of lane choice to evaluate alternative HOT lane pricing strategies, including the use of GP density, to more accurately reflect the value of the HOT lane. In addition, the paper explores the potential effect these strategies would have on the HOT lane vehicle share through a partial equilibrium analysis. This analysis demonstrates the change in demand elasticity with price, showing the point at which drivers switch from a positive to negative elasticity.

While toll roads originated to finance infrastructure and earn profits for their owners, it has been long understood that tolls could also be used to manage traffic, as higher tolls tend to deter usage (*1–5*). The construction of high occupancy vehicle (HOV) lanes, beginning with the Shirley Highway in 1969, launched as a bus-only lane and subsequently opened to carpools in 1973 during the energy crisis, aimed to provide an incentive for travelers to carpool to increase average auto occupancy and person-throughput on roadways, and thus lower energy use. In general however, such facilities were inefficient, either the occupancy threshold was too low, and the lanes carried as many vehicles as general purpose (GP) lanes, or it was too high, and the lane was underutilized, leaving vehicle capacity unused (*6*, *7*).

In recent years many HOV lanes in the United States have been converted to high occupancy/toll (HOT) lanes, opening the lanes which were underused to additional toll-paying traffic. The toll serves to raise revenue to cover operating costs and to regulate the demand of single occupant vehicles (SOVs). HOT lanes across the United States use different methods for determining the toll; however, all such facilities raise the toll during more congested periods. The theory is, a higher toll price discourages demand and is used to maintain a high level of service in the HOT lane(s). In practice, however, Janson and Levinson (*8*) showed that a higher price may act as a signal of downstream congestion (in both the GP and HOT lanes), causing demand for the HOT lane to increase, at least to a point. Evaluations of alternative pricing schemes have been limited (*9*).

This paper develops a model to predict HOT lane choice on Minnesota's well-studied MnPASS lanes (*10–15*). It then develops some alternative HOT pricing strategies. These alternative strategies are tested using a partial equilibrium analysis. This analysis uses the calibrated HOT lane choice model to predict the HOT lane share at various prices and estimate demand elasticity to price.

## Pricing on MnPASS Lanes

The MnPASS lanes in Minneapolis were an early application of dynamic toll pricing in the United States, opening in 2005. With several exceptions, the general operating hours are 06:00–10:00 and 15:00–19:00. Prices during operation range from a minimum of \$ 0.25 to a maximum \$8.00. Each route (I-394, I-35W, and I-35E) is divided into multiple sections with prices posted for use of each segment. The maximum price applies to use of each section individually, as well as use of all sections.

[1]SRF Consulting Group, Minneapolis, MN
[2]School of Civil Engineering, University of Sydney, Sydney, New South Wales, Australia

**Corresponding Author:**
Address correspondence to David Levinson: david.levinson@sydney.edu.au

**Table 1.** Pricing Plan for Normal Operation of MnPASS Lanes (both I-35W and I-394) at Time of Study

| Level of service | Min density | Max density | Min rate ($) | Default rate ($) | | Max rate ($) |
|---|---|---|---|---|---|---|
| A | 0 | 11 | 0.25 | 0.25 | | 0.50 |
| B | 12 | 18 | 0.50 | 0.50 | | 1.50 |
| C | 19 | 31 | 1.50 | 1.50 | | 2.50 |
| D | 32 | 42 | 2.50 | 3.00 | | 3.50 |
| E | 43 | 49 | 3.50 | 5.00 | | 5.00 |
| F | 50 | 50 | 5.00 | 8.00 | | 8.00 |
| Change in price from density change | | | | | | |
| Density | Δ 1 | Δ 2 | Δ 3 | Δ 4 | Δ 5 | Δ 6 |
| 0-18 | 0.25 | 0.25 | 0.25 | 0.25 | 0.25 | 0.25 |
| 19 + | 0.25 | 0.50 | 0.75 | 1.00 | 1.25 | 1.50 |

*Note:* Density (*K*) in veh/mi/ln; Prices in $.

Prices are adjusted every three minutes based on density levels measured in the MnPASS lanes only. Traffic levels in the GP lanes do not directly influence price. Loop detector counts are taken every 30 s and used to calculate the density in the MnPASS virtual toll collection plazas along the corridor. Density measurements are averaged over the most recently completed six-minute period in order to smooth out fluctuations and based only on downstream congestion. Price is dictated by the magnitude of density as well as the change in density over the previous six minutes. A rise in density creates an increase in price.

Table 1 displays the pricing plan in effect for most of the history of the express lanes, and the plan in effect when the empirical analyses described below were conducted. It remained in effect until 2015, when it was replaced by a continuous function (*16*). Both of these approaches regulate the price based on traffic density level. Minimums and maximums for a given level of service must be maintained. The table also indicates the changes in price caused by a change in density.

## HOT Lane Choice

### *Methods*

The analysis uses a fixed number of SOVs with predefined commute times and locations to calibrate a lane choice model. Transponder-equipped SOVs can decide whether to use the MnPASS or GP lanes based on the toll and their expected travel time and reliability.

The HOT lane choice model extends Carrion (*17*), which addressed an earlier version of the question with a more limited dataset. The binomial logit model determines the probability of a vehicle using the HOT lane based on several independent variables. These variables include estimated travel time and travel time variability for both the HOT lane and the GP lanes, as well as the posted toll price. The lane choice model applies only to SOVs equipped with transponders. SOVs not equipped with transponders are not allowed to use the MnPASS lanes. A separate subscription choice model was developed to determine which vehicles are equipped with transponders (*18*).

Utility of a lane is described as

$$U = f(T, V, P, A)$$

where
T = expected travel time,
V = travel time variability,
P = expected toll price, and
A = alternative specific constant.

The utility decreases with an increase in expected travel time, decreasing the probability of using the given lane type. Expected travel time is measured in minutes. This value is computed for HOT lanes and GP lanes separately.

Travel time variability in this model is defined as the 90th percentile–50th percentile to correspond with Carrion (*17*). This value is calculated separately for the HOT lane and GP lanes. Like expected travel time, an increase in variability decreases the probability of using that lane. Travel time variability is measured in minutes.

The expected toll price variable is based on the dynamic message sign posted price. The price corresponds to a user's entry and exit points. This model assumes all drivers will exit in downtown Minneapolis. Therefore, the expected toll will vary only by entry point. Toll prices are in USD. The negative sign indicates a dissuasion from higher tolls, assuming all other factors remain constant.

In this model, the alternative specific constant (ASC) was defaulted to zero and adjusted if necessary in the calibration.

The lane choice model was calibrated by matching a set of simulated vehicles' HOT lane decisions to historical data. A list of vehicles was generated from trip tables provided by the Metropolitan Council. All vehicles are

SOVs traveling eastbound to downtown Minneapolis on I-394 between 06:00 and 10:00. Each vehicle has an entrance ramp and time of entry into the system. The subscription choice model from Owen et al. (*18*) is first applied to filter non-transponder-owning SOVs. Each vehicle experiences various travel times based on the entrance ramp and time of entry. These travel times are the basis of the expected travel time and travel time reliability parameters of the lane choice model. Details of the calibration steps are outlined below.

The lane choice model coefficients are adjusted using a grid search technique. Default values for the coefficients were taken from Carrion (*17*), with the exception of the ASC which was set to zero. The grid search approach involves adjusting each of the coefficients separately (travel time first, then tolls). The first coefficient is altered until the model achieves its best fit to the calibration target. This coefficient is then kept constant and the second coefficient is adjusted until the fit can no longer be improved. The ratio of expected travel time to travel time variability was kept constant due to the extensive literature research in determining this value outlined in Carrion and Levinson (*19*).

Each vehicle builds a travel time history by experiencing MnPASS travel times along the corridor based on their entrance ramp and time of entry. All travel is along I-394 Eastbound to downtown Minneapolis. The travel times are calculated using loop detector data from each Wednesday of 2012 except holidays: July 4 and December 26. This travel history determines a vehicle's expected travel time (mean of travel history) and travel time variability (90th percentile minus 50th percentile).

In order to calibrate the lane choice model, it is necessary to determine the probability that a transponder-owning SOV will use the MnPASS lane.

Using Bayes' theorem

$$\Pr(L|R) = \Pr(R|L)*\Pr(L)/\Pr(R) \quad (1)$$

$\Pr(R)$ is the probability of radio transponder ownership (from subscription choice model). $\Pr(L)$ represents the probability of using the HOT lane among all SOVs. $\Pr(R|L)$ is the probability of owning a transponder given use of the HOT lane. Since only SOVs are being considered, $\Pr(R|L)$ is 1 (or 100%) assuming no illegal use of the HOT lane. The rate of non-compliance is unknown, but estimated to be 10%–20% (*20*). Vehicle violations do not affect the model in a significant way, as they are instead assumed to be HOVs, which are also not directly measured. Violations do vary by section. One section of I-394 is barrier-separated, the remainder of the system is more open, separated from GP lanes by dashed lines, (I-35W) or painted double white lines with periodic dashed lines (I-394, I-35E) to organize entry and exit into the HOT lane system.

$\Pr(L)$ was calculated by finding the number of SOVs using the MnPASS lane and dividing by total number of vehicles using the corridor during the same time period. Total vehicle counts were gathered from loop detector data. The number of HOVs using the GP lanes is assumed to be zero. Counts of SOVs using the MnPASS lane come from transponder data which shows entry and exit plazas and entry time, along with paid toll price. By comparing the counts throughout morning peak period with the GP loop detector data, $\Pr(L)$ can be determined.

$\Pr(R)$ was calculated by correlating the subscription choice model in Owen et al. (*18*) with subscription data for each transportation analysis zone (TAZ) along the corridor. Each vehicle's entrance ramp can be probabilistically correlated to surrounding TAZs. By then applying the subscription choice model to the total set of SOVs, a subset of transponder-equipped SOVs is formed.

Due to the connection of this research to the calibrated traffic simulation, November 29, 2011 was selected for calibration of the lane choice model. It was an average day with no weather or crash-related problems along the MnPASS corridors. Any specific day has variable demand and driving changes due to daylight, temperature, visibility etc., and future research should calibrate to more days to mitigate this, but we did not observe anything out of the ordinary on the day.

The $\Pr(L)$ value from that day and $\Pr(R)$, result in

$$\Pr(L|R) = (100\%)*(11.8\%)/(17.3\%) = 68.1\% \quad (2)$$

Although the MnPASS toll price fluctuates based on HOT density, there is a direct correlation between the toll and the time savings the MnPASS lanes provide over the GP lanes; so the higher the toll, the greater the time savings. This correlation is observed by users and explains the positive demand elasticity to price results described in Janson and Levinson (*8*).

### Results

Using average toll prices and time savings data from 2012, a log relationship was fit. The bimodal relationship of the data meant two log functions were fit, one for congestion onset and one for offset.

The relationship between price and time savings (and travel time variance) during congestion onset and offset are displayed in Table 2. The corresponding price time-savings curves are displayed in Figure 1. The price reliability curves are displayed in Figure 2.

### Calibration Process

The flowchart (Figure 3) displays the lane choice model calibration cycle using the grid search technique. Once lane choice decisions for all vehicles have been completed, the percentage of vehicles using the HOT lane

**Table 2.** Price-Time Savings and Price-Reliability Regression Results

| Variable | Time savings vs. price | | Time variance difference vs. price | |
|---|---|---|---|---|
| | Onset | Offset | Onset | Offset |
| Intercept | 0.5527*** | 1.2965*** | 0.9566*** | 1.6636*** |
| | (0.05972) | (0.02912) | (0.03867) | (0.01664) |
| *ln(P)* | 1.2587*** | 0.7953*** | 1.1413*** | 0.926*** |
| | (0.07732) | (0.03743) | (0.05006) | (0.02139) |
| *n* | 40 | 40 | 40 | 40 |
| $r^2$ | 0.8923 | 0.913 | 0.942 | 0.9657 |

*Note*: (Standard error in parentheses).
Significance *0.05, **0.01, ***0.001.
Time Savings and Time Variance Difference in minutes are the dependent variables, price in USD is the independent variable.

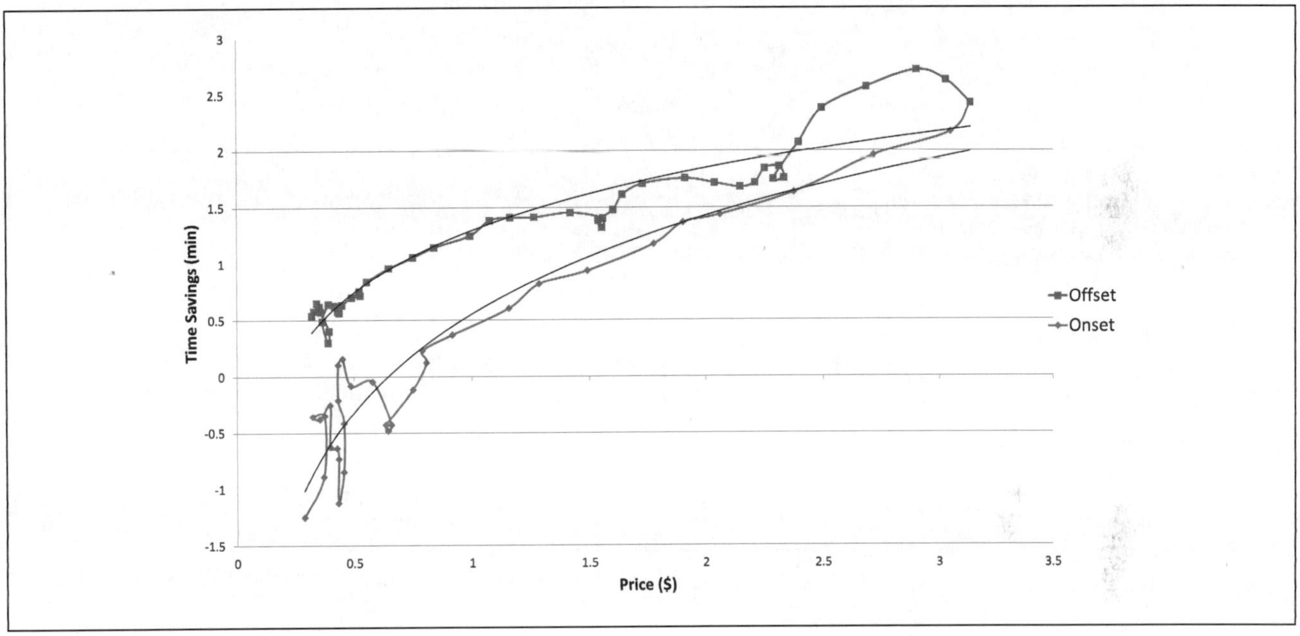


**Figure 1.** Price-time savings log model.
*Note*: $\Delta T_{onset} = 1.2587 \ln(P) + 0.5527$ ($r^2 = 0.8923$).
$\Delta T_{offset} = 0.7953 \ln(P) + 1.2965$ ($r^2 = 0.913$).
where ΔT is travel time savings in minutes and P is price in USD.

($\Pr(L|R)$) is compared to the calibration target of 68.1%. The model coefficients are then adjusted to increase or decrease HOT lane share and the process is repeated until the optimal coefficients are found.

The lane choice parameters were for both congestion onset (at the beginning of the peak moving from free-flow to congested conditions) and offset (the end of the peak, restoring freeflow from congested conditions). The resulting values are found in Table 3.

## Alternative Pricing Scenarios

Results from the lane choice model are used to test four alternative pricing strategies using a fixed demand partial equilibrium analysis.

Four HOT lane pricing strategies are tested that could serve as alternatives to the previous MnPASS pricing system. The previous system relied on a series of density and price tables to determine the toll based strictly on HOT lane density. The proposed alternatives determine the toll based on a mathematical function relating HOT lane density (and GP lane density in three of the strategies) to price. The three value pricing strategies use the difference in GP and HOT lane density to determine the toll. Due to the nonlinear relationship between density and time savings, two of the strategies are weighted by either HOT density or GP density.

The continuous function is similar to the previous pricing algorithm in that it relies strictly on HOT lane density for determining price; however, instead of relying

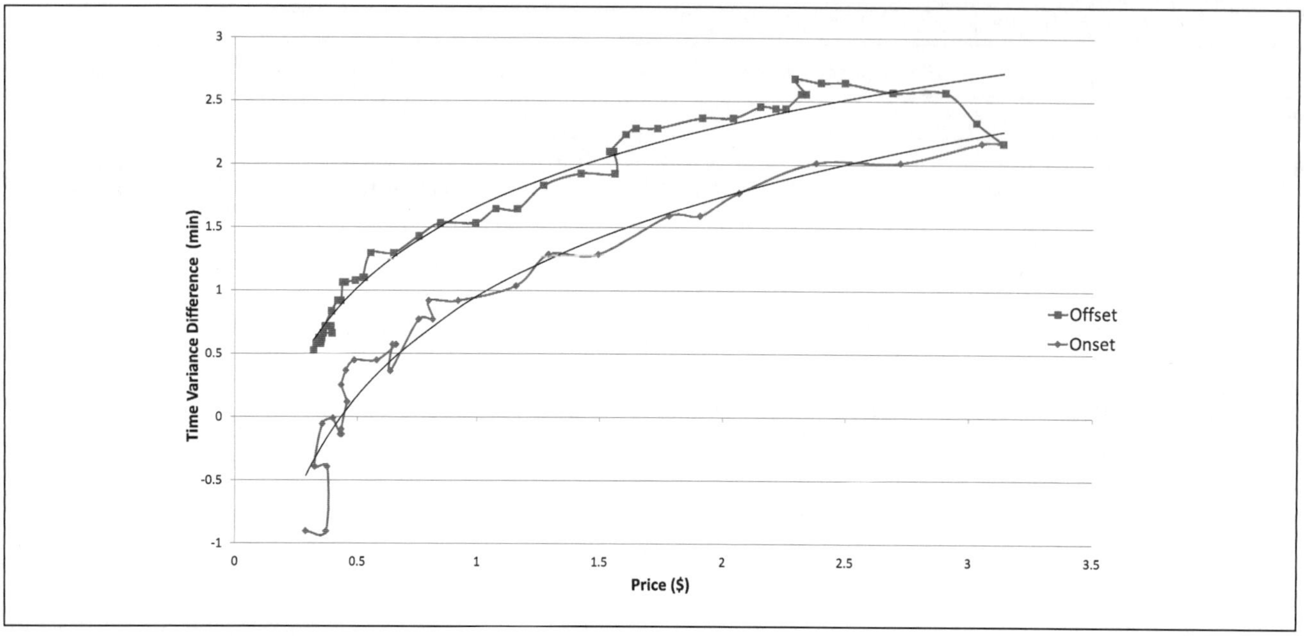


**Figure 2.** Price-reliability model.
*Note:* $\Delta V_{onset} = 1.1413\ln(P) + 0.9566$ ($r^2 = 0.942$).
$\Delta V_{offset} = 0.926\ln(P) + 1.6636$ ($r^2 = 0.9657$).
where $\Delta V$ is time variance difference in minutes and P is price in USD.

**Table 3.** Lane Choice Model Parameters for Calibration

| Parameter | Carrion (*17*) | Onset | Offset |
|---|---|---|---|
| Expected travel time | −0.672 | −7.27 | −10.7 |
| Travel time variability | −0.228 | −2.47 | −3.63 |
| HOT lane toll | −6.94 | −6.94 | −6.94 |
| Alternative specific constant | −2.23 | 0 | 0 |

**Table 4.** Pricing Strategy Coefficients for Constant Time Savings Lane Choice Model

| Pricing strategy | Iteration | Coefficient | Lane share | Average price |
|---|---|---|---|---|
| Continuous Function Pricing ($\alpha$): | 1 | 0.059 | 78.3% | $1.77 |
| (where $\beta$ = 1.156) | 2 | 0.085 | 68.1% | $2.39 |
| Unweighted Value Pricing ($\gamma$): | 1 | 0.058 | 65.8% | $2.16 |
| | 2 | 0.051 | 68.1% | $1.87 |
| $HOT_{Weighted}$ Value Pricing ($\delta$): | 1 | 0.0034 | 67.0% | $2.77 |
| | 2 | 0.0033 | 68.1% | $2.51 |
| $GP_{Weighted}$ Value Pricing ($\sigma$): | 1 | 0.0017 | 58.3% | $3.45 |
| | 2 | 0.00089 | 68.1% | $2.28 |

on a series of tables, price is determined from a mathematical equation. The three other value pricing strategies incorporate GP lane density and use the difference in density between the HOT and GP lanes to determine price. Details of the pricing strategies are outlined below.

In all cases, the prices are confined to several constraints to match the existing pricing algorithm. The minimum price is $ 0.25, the maximum $8.00 and all prices are rounded to the nearest $ 0.25. The following equation represents the constraints which are applied after the unconstrained price is determined.

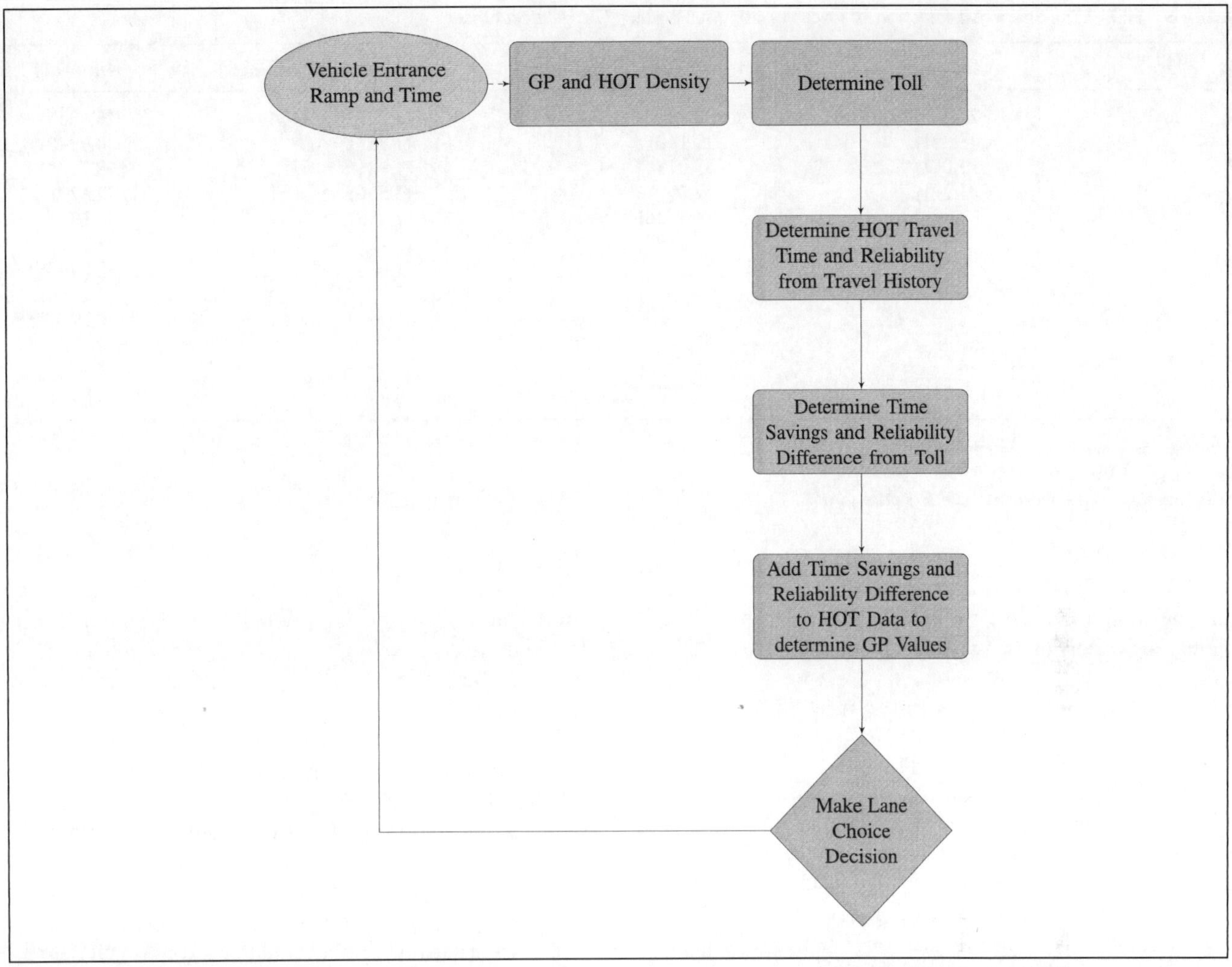


**Figure 3.** Calibration cycle.

$$P_{\text{constrained}} = \text{Rnd}(\text{Min}(\text{Max}(P_{\text{unconstrained}}, 0.25), 8.00), 0.25) \quad (3)$$

$P_{\text{unconstrained}}$ may be defined several ways, as discussed below.

Continuous Function: Prices using this function are determined by

$$P_{\text{continuous}} = \alpha^* K_{\text{HOT}}^{\beta} \quad (4)$$

where P represents the price in USD and K the density in vehicles/mile/lane.

$K_{\text{HOT}}$ is found using the same method as the current algorithm (maximum downstream density averaged over last 6 min). $\alpha$ and $\beta$ are constants which can be adjusted to achieve the desired curve.

Unweighted Value Pricing: While the current pricing algorithm only evaluates the density in the HOT lane, this pricing strategy computes price based on the difference in density between the GP and HOT lanes. The difference in density between the lane groups is correlated with a difference in time savings and therefore, the value provided by the HOT lane. Implementation of this pricing scheme (and subsequent strategies), will require the integration of GP density as a factor in determining price. GP density is averaged among parallel detectors. The maximum downstream GP density is then used to determine price, along with the maximum downstream HOT density.

$$P_{\text{Value}_{\text{unweighted}}} = \gamma^*[K_{\text{GP}} - K_{\text{HOT}}] \quad (5)$$

$\text{HOT}_{\text{weighted}}$ Value Pricing: Differences in density between GP and HOT lanes do not correlate directly to travel speeds. Rather, there is a correlation with the magnitude of densities. For example, little speed difference exists between 10 and 20 vehicles/mi/ln (6 and 12 veh/km/ln), both likely experience free flow speeds. However, a greater speed difference exists at higher densities (between 40 and 50 veh/mi/ln (approximately 25 and 31

**Table 5.** HOT Lane Share as a Function of Price: Continuous Pricing, Congestion Onset

| Variable | 1st order | 2nd order | 3rd order | 4th order |
|---|---|---|---|---|
| Intercept | 21.46*** | −0.9280 | −15.297*** | −25.884*** |
| | (5.321) | (3.140) | (1.997) | (0.9949) |
| P | 1.017 | 35.019*** | 71.931*** | 108.91*** |
| | (1.291) | (2.892) | (3.546) | (2.670) |
| $P^2$ | – | −4.5576*** | −17.218*** | −40.396*** |
| | – | (0.3788) | (1.129) | (1.516) |
| $P^3$ | – | – | 1.0781*** | 5.7446*** |
| | – | – | (0.09485) | (0.2956) |
| $P^4$ | – | – | – | −0.2941*** |
| | – | – | – | (0.01849) |
| n | 44 | 44 | 44 | 44 |
| $r^2$ | 0.0146 | 0.7825 | 0.9486 | 0.9931 |

*Note*: (Standard error in parentheses).
Significance * 0.05, ** 0.01, *** 0.001.
HOT lane share is dependent variable, P is price in USD.

veh/km/ln)). This function weights the difference based on the magnitude of the HOT lane density. Price increases proportionally with HOT density.

$$P_{\text{ValueHOT}_{\text{weighted}}} = \delta^*[K_{GP} - K_{HOT}]^*K_{HOT} \quad (6)$$

$GP_{\text{weighted}}$ Value Pricing: This pricing strategy is weighted based on GP density instead of HOT density. If $K_{GP}$ is much greater than $K_{HOT}$ and $K_{HOT}$ is very low, then the HOT weighted value pricing strategy would yield a low price even though there would be a significant value in using the HOT lane. By weighting based on $K_{GP}$, this strategy ties price more directly to the GP lane congestion and the actual time savings gained by using the HOT lane.

$$P_{\text{ValueGP}_{\text{weighted}}} = \sigma^*[K_{GP} - K_{HOT}]^*K_{GP} \quad (7)$$

The $\text{HOT}_{\text{weighted}}$ strategy combines the value pricing concept with the current algorithm's direct correlation between HOT density and price. For this reason, we posit that this pricing strategy would provide the greatest improvement over the current pricing system while still maintaining some of the same logic. The continuous function, on the other hand, most closely resembles the previous pricing system (and was subsequently implemented), but does not account for the density in the GP lanes.

## *Evaluation*

The calibrated HOT lane choice model is used to test the behavior of the alternative pricing strategies and how changing prices affect the share of transponder-owning SOVs that use the MnPASS lane.

Each pricing strategy's coefficients were incrementally adjusted and the process rerun to determine the resulting HOT lane share ($\Pr(L|R)$). The average price and share were recorded for each iteration. The results were graphed and fit for each pricing strategy (congestion onset and offset).

The coefficients are shown in Table 4. Each line represents an iteration.

The first iteration uses parameters to best match the previous algorithm based on average demand data from 2011 and 2012.

The second iteration adjusts parameters to best match the HOT lane share calibration target of 68.1%.

Table 5 displays the regression results from fitting the continuous pricing strategy using a first, second, third, and fourth order polynomial function. The fourth degree polynomial functions for each scenario are displayed in Table 6 and graphs of the continuous function (congestion onset and offset) are displayed in Figure 4.

## *Elasticity*

The functions above describe HOT lane share as a function of toll price. The elasticity of share to price is determined by taking the derivative of the function and multiplying by the quotient of price divided by share.

$$\varepsilon_{\Pr(L|R)(P)} = \frac{P^*\Pr(L|R)'(P)}{\Pr(L|R)(P)} = \frac{d\ln\Pr(L|R)(P)}{d\ln P} \quad (8)$$

Figure 5 graphs elasticity as a function of price for the continuous function pricing strategy (onset and offset). The elasticity equations are displayed below the figure.

Average elasticity values are shown in Table 7.

## *Discussion of Results*

All four pricing strategies show a similar pattern in the relationship between HOT lane share and price. The

**Table 6.** Pricing Function Model Equations

| Pricing function | Model | $r^2$ |
|---|---|---|
| Onset | | |
| Continuous | $Pr(L\|R) = -0.2941P^4 + 5.7446P^3 - 40.396P^2 + 108.91P - 25.884$ | 0.9931 |
| Unweighted | $Pr(L\|R) = -0.1555P^4 + 3.6124P^3 - 31.812P^2 + 106.14P - 26.573$ | 0.9909 |
| $HOT_{weighted}$ | $Pr(L\|R) = -0.3468P^4 + 6.3111P^3 - 40.349P^2 + 98.579P - 22.116$ | 0.9493 |
| $GP_{weighted}$ | $Pr(L\|R) = -0.1785P^4 + 3.2471P^3 - 22.649P^2 + 66.301P - 13.515$ | 0.9604 |
| Offset | | |
| Continuous | $Pr(L\|R) = -0.2394P^4 + 4.4049P^3 - 27.423P^2 + 51.688P + 61.127$ | 0.9904 |
| UnweightedValue | $Pr(L\|R) = -0.1284P^4 + 3.0066P^3 - 23.308P^2 + 51.99P + 61.762$ | 0.9887 |
| $HOT_{weighted}$ | $Pr(L\|R) = -0.1652P^4 + 2.9438P^3 - 17.877P^2 + 29.923P + 69.300$ | 0.9801 |
| $GP_{weighted}$ | $Pr(L\|R) = -0.0546P^4 + 1.2879P^3 - 10.691P^2 + 22.483P + 68.142$ | 0.9905 |

*Note:* P is price in USD.

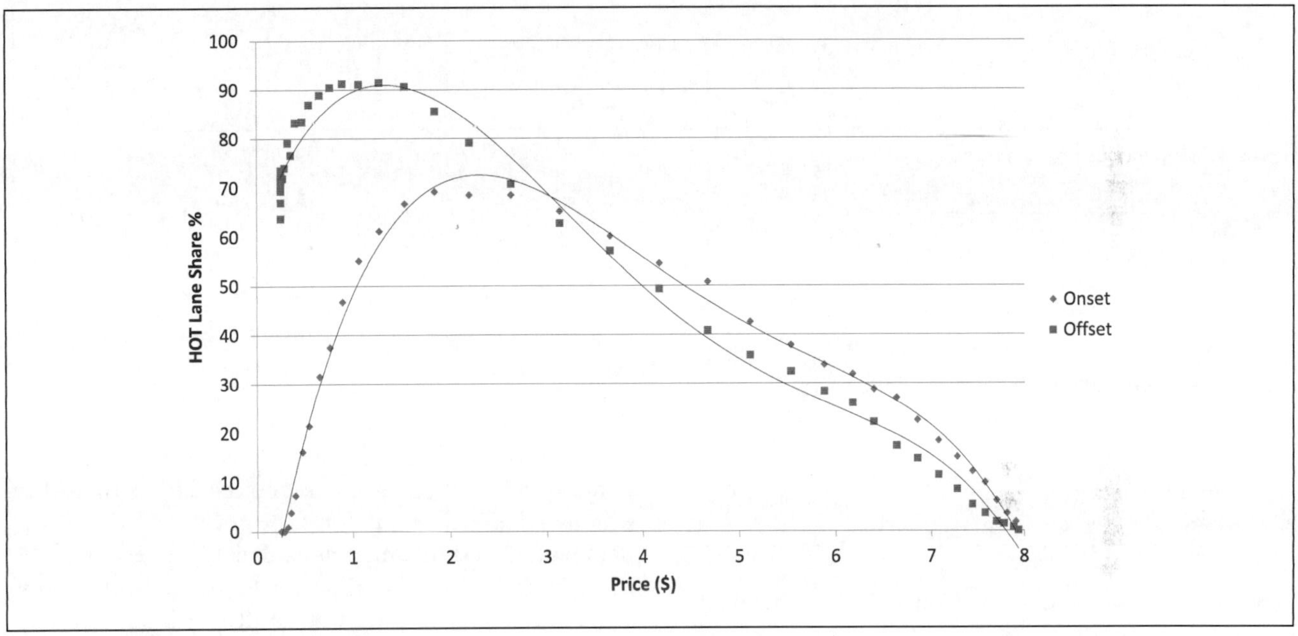


**Figure 4.** Continuous pricing function.
$P = \alpha * K_{HOT}^{\beta}$, where $\beta$=1.156 and $\alpha$ = 0.065.

maximum HOT lane share during congestion onset is achieved between \$2 and \$3, whereas during congestion offset, the greatest HOT lane share occurs between \$1 and \$2. In general, the HOT lane share during congestion offset is greater than during the onset due to the greater time savings and reliability per dollar of toll as demonstrated previously in Figures 1 and 2. Table 8 shows the average price and HOT lane share for each pricing strategy along with the standard deviation.

Figure 4 shows the rise and fall of the HOT lane share as the toll increases (and therefore, time savings). When HOT lane share reaches its maximum, elasticity switches from positive to negative. Figure 6 outlines how changes in toll and time savings affect HOT lane share and ultimately, elasticity to price.

At lower tolls, an increase in price results in a higher HOT lane share (positive elasticity), whereas at higher tolls, an increase in price causes a decrease in HOT lane share (negative elasticity). At lower tolls, the improved time savings and reliability outweigh the increase in toll. However, at higher tolls, the increase in toll outweighs greater time savings and reliability causing the HOT lane share to decrease. This is consistent with the idea that, after a certain point, each marginal minute of travel time saving is of lower value. Future research can infer how value of time varies with congestion level, and if there is a maximum willingness-to-pay.

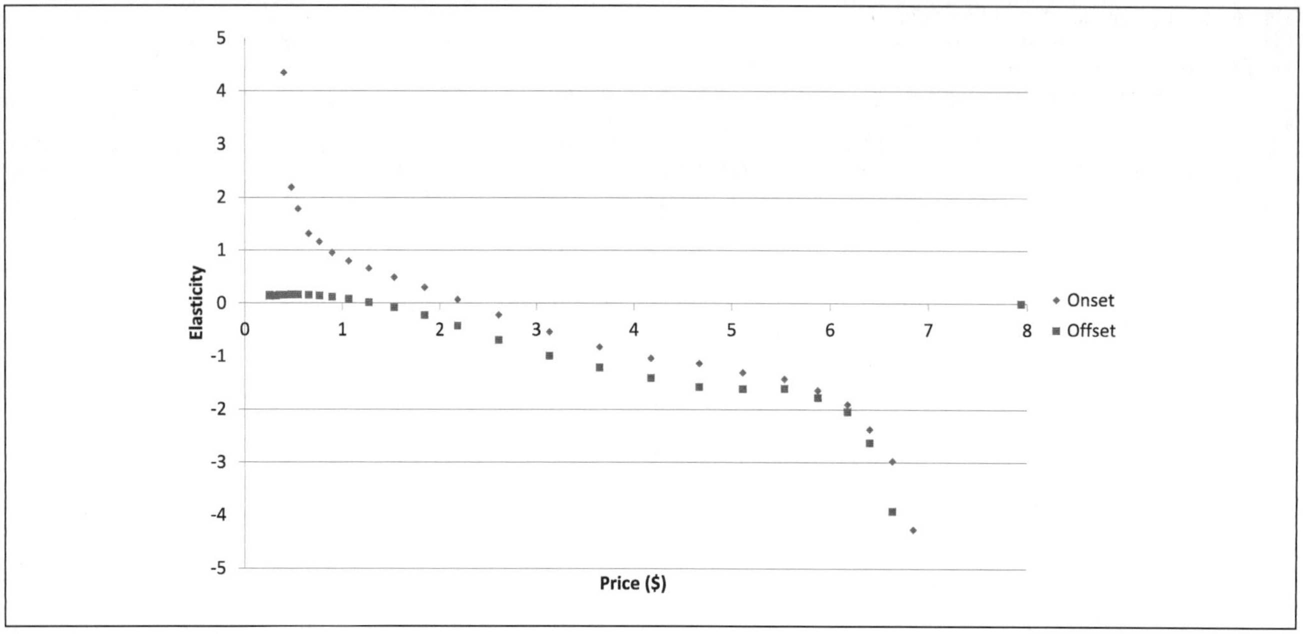


**Figure 5.** Continuous pricing function.

*Note*: $\varepsilon_{\Pr(L|R)(P)} = \frac{P*(-1.176*4*P^3 + 17.23*P^2 - 80.79*P + 108.9)}{\Pr(L|R)(P)}$

$\varepsilon_{\Pr(L|R)(P)} = \frac{P*(-0.9576*P^3 + 13.21*P^2 - 54.85*P + 51.69)}{\Pr(L|R)(P)}$

**Table 7.** Average Elasticity Values (Fixed Time Savings)

| | |
|---|---|
| Continuous pricing | −0.372 |
| Unweighted value pricing | −0.260 |
| HOT weighted value pricing | −0.175 |
| GP weighted value pricing | −0.496 |

Based on an earlier version of this research, Minnesota Department of Transportation (MnDOT) selected the continuous function to replace the toll table described in Table 6.

## Conclusion

This research estimated and calibrated a lane choice model for HOT lanes. In the model, simulated vehicles made decisions on whether to use the MnPASS lane based on the toll and their anticipated time savings and improved travel time reliability. Four alternative pricing scenarios were tested. The lane choice was determined at various price increments for each pricing system. These were plotted and fit with a polynomial, the derivatives of which correlate to the elasticity to price. In all cases, demand elasticity to price was positive at lower tolls and negative at higher tolls. MnPASS users recognize the correlation between the toll price and the time savings and travel time reliability provided by the lanes. The toll price acts as a proxy of downstream congestion. At lower tolls, the travel time savings and reliability advantages outweigh the cost of the toll and share of HOT lane users rises. However, at higher tolls, the cost of using the lane begins to outweigh the benefit and the share drops.

**Table 8.** Average HOT Lane Share and Prices

| Pricing strategy | Avg price ($) | Std dev price ($) | Avg HOT lane share (%) | | Std dev HOT lane share (%) | |
|---|---|---|---|---|---|---|
| | | | Onset | Offset | Onset | Offset |
| Continuous | 2.93 | 2.93 | 54.6 | 24.4 | 31.6 | 24.7 |
| Unweighted | 3.19 | 3.20 | 54.1 | 24.5 | 36.9 | 32.6 |
| HOT weighted | 3.65 | 2.93 | 49.4 | 24.5 | 31.6 | 19.3 |
| GP weighted | 3.83 | 3.26 | 45.5 | 18.9 | 34.0 | 21.0 |

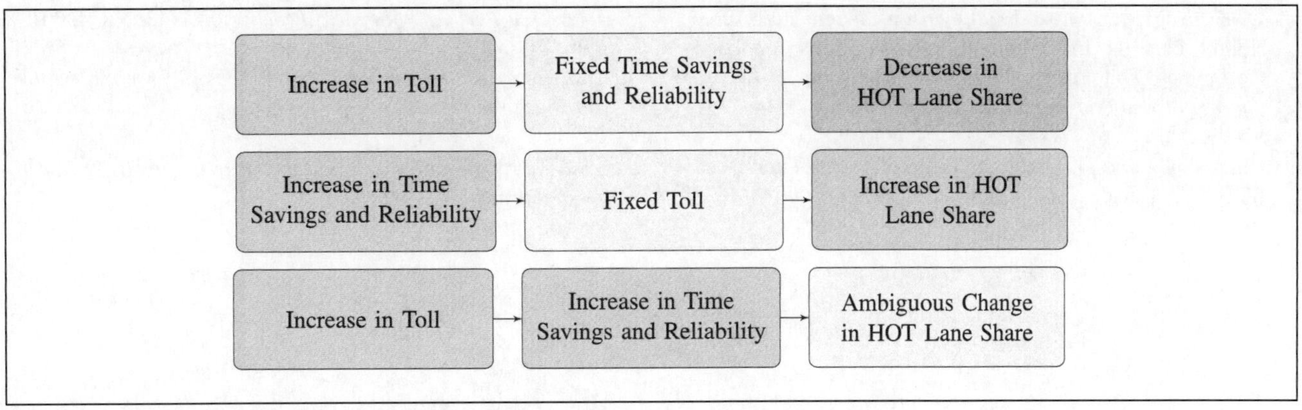


**Figure 6.** Toll and time savings effect on HOT lane share.

These results are estimated on a system where drivers have incomplete information about travel time savings from HOT lane usage, and use price as a signal of time savings. In a context where drivers were better informed (e.g. through variable message signs or real-time congestion-aware GPS navigation systems), results would likely be significantly different.

Future research should field-test alternative pricing strategies and parameter values to identify which best achieves the goals of maximizing use of the HOT lanes while maintaining reliable free-flowing speeds, recognizing that travelers may change their sensitivity to price if the perceived relationship between price and travel time savings changes.

## Author Contributions

The authors confirm contribution to the paper as follows: study conception and design: DL, MJ; data collection: MJ; analysis and interpretation of results: DL, MJ; draft manuscript preparation: DL, MJ. All authors reviewed the results and approved the final version of the manuscript.

## References

1. Dupuit, J. On the Measurement of the Utility of Public Works. *International Economic Papers*, Vol. 2, No. 1952, 1844, pp. 83–110.
2. Vickrey, W. S. Congestion Theory and Transport Investment. *The American Economic Review*, Vol. 59, No. 2, 1969, pp. 251–260.
3. Pahaut, S., and C. Sikow. History of Thought and Prospects for Road Pricing. *Transport Policy*, Vol. 13, No. 2, 2006, pp. 173–176.
4. Yang, H., and H.-J. Huang. *Mathematical and Economic Theory of Road Pricing*. Oxford, UK: Elsevier, 2005.
5. Small, K. A., C. Winston, and C. A. Evans. *Road Work: A New Highway Pricing and Investment Policy*. Washington, DC: Brookings Institution Press, 2012.
6. Dahlgren, J. High Occupancy Vehicle Lanes: Not Always More Effective than General Purpose Lanes. *Transportation Research Part A: Policy and Practice*, Vol. 32, No. 2, 1998, pp. 99–114.
7. Dahlgren, J. High-occupancy/Toll Lanes: Where Should They be Implemented? *Transportation Research Part A: Policy and Practice*, Vol. 36, No. 3, 2002, pp. 239–255.
8. Janson, M., and D. Levinson. HOT or Not: Driver Elasticity to Price on the MnPASS HOT Lanes. *Research in Transportation Economics*, Vol. 44, 2014, pp. 21–32.
9. Gardner, L., S. Boyles, H. Bar-Gera, and K. Tang, Robust Tolling Schemes for High-Occupancy Toll Facilities Under Variable Demand. *Transportation Research Record: Journal of the Transportation Research Board*, 2014. 2450: 152–162.
10. Halvorson, R., and K. R. Buckeye, High-occupancy Toll Lane Innovations: I-394 MnPASS. *Public Works Management & Policy*, Vol. 10, No. 3, 2006, pp. 242–255.
11. Munnich, L., and K. Buckeye, I-394 MnPASS High-Occupancy Toll Lanes: Planning and Operational Issues and Outcomes (Lessons Learned in Year 1). *Transportation Research Record: Journal of the Transportation Research Board*, 2007. 1996: 49–57.
12. Tilahun, N., and D. Levinson. Unexpected Delay and the Cost of Lateness on I-394 High Occupancy/Toll Lanes. *Travel Demand Management and Road User Pricing: Success, Failure and Feasibility*, 2009, pp. 173–184.
13. Goodall, N., and B. Smith, What Drives Decisions of Single-occupant Travelers in High–occupancy Vehicle Lanes? Investigation Using Archived Traffic and Tolling Data from Mn-PASS Express Lanes. *Transportation Research Record: Journal of the Transportation Research Board*, 2010. 2178: 156–161.
14. Cao, X. J., L. Munnich, H. Liu, X. S. He, Z. Xu, and Y. A. Huang. *Benefit and Cost Analysis of the I-394 MnPASS Program*, 2012.
15. Buckeye, K. Performance Evaluation of I-394 MnPASS Express Lanes in Minnesota. *Transportation Research Record: Journal of the Transportation Research Board*, 2012. 2278: 153–162.
16. Harlow, T. The Drive: MnDOT to Tweak MnPass pricing. *Star Tribune*, November 9, 2015, 2015.
17. Carrion, C. *Value of Reliability: Actual Commute Experience Revealed Preference Approach*. Ph.D. thesis, University of Minnesota, 2010.

18. Owen, A. M. Janson, and D. Levinson. Incremental Accessibility Benefits and Choice of Subscriptions for High-Occupancy Toll Lanes. *Transportation Research Record: Journal of the Transportation Research Board*, 2014. 2412: 93–99.
19. Carrion, C., and D. Levinson. Value of Travel Time Reliability: A Review of Current Evidence. *Transportation Research Part A: Policy and Practice*, Vol. 46, No. 4, 2012, pp. 720–741.
20. Raddatz, K. *Infrared Camera Helping Troopers Catch MnPass Violators*. WCCO, 2017.

*The Standing Committee on Congestion Pricing (ABE25) peer-reviewed this paper (18-00376).*

Article

TRR
JOURNAL OF THE TRANSPORTATION RESEARCH BOARD

Transportation Research Record
2018, Vol. 2672(5) 23–32
© National Academy of Sciences: Transportation Research Board 2018
Article reuse guidelines: sagepub.com/journals-permissions
DOI: 10.1177/0361198118796928
journals.sagepub.com/home/trr
SAGE

# Unrevealed Preferences: Unexpected Traveler Response to Pricing on Managed Lanes

**Mark W. Burris[1] and John F. Brady[2]**

**Abstract**

This paper addresses priced managed lane corridors, on which travelers may choose to pay a toll to travel on the managed lanes (MLs) to realize generally faster, more reliable travel than on the adjacent, toll-free general purpose lanes (GPLs). These lanes exist in many cities across the United States and are becoming more common as transportation agencies look for innovative ways to increase capacity and regulate demand for their roadways. Commonly, demand for these lanes is modeled assuming travelers choose between the MLs and GPLs primarily based on the cost and time savings of the MLs. Although the traffic and revenue forecasts generated by these models have generally succeeded in estimating revenue, newly available empirical data from Katy Freeway and North Tarrant Express shows these models fail to capture how individual drivers make decisions. Most travelers on those freeways were not choosing—they always used the same lane regardless of travel time and toll. Travelers that used both sets of lanes often made choices that appeared counter-intuitive based on travel time savings and toll rate. This research provides a preliminary investigation into this issue, which calls into question all prior ML travel-behavior research.

Priced managed lanes (MLs) present travelers with the option to pay a toll or meet certain other requirements to travel on the generally faster MLs. These lanes exist in many cities across the United States and are becoming more common as transportation agencies look for innovative ways to increase capacity and regulate demand for their roadways. The most common form of priced MLs is high occupancy/toll (HOT) lanes, on which high occupancy vehicles (HOVs) can travel for free or a reduced price but single occupancy vehicles (SOVs) must pay a toll—examples include I-15 in San Diego, I-35 in Minnesota, I-85 in Atlanta, I-495 in Northern Virginia, Katy Freeway in Houston, and North Tarrant Express (NTE) in Dallas–Fort Worth.

On a number of dimensions MLs have been successful: I-495, I-95 in Florida, and NTE have all reported substantial increases in throughput and decreases in corridor congestion following the addition of MLs (*1*). The revenue generated by priced MLs supports numerous initiatives, usually being reinvested in the corridor upkeep, supporting alternative transportation options, and repaying the cost of construction.

By design, MLs present users of a corridor with a simple choice: choose the generally slower-moving general purpose lanes (GPLs) or pay a toll to travel on the generally faster MLs. Understanding how drivers value, compare, and choose between these two alternatives now and in the future is a critical consideration in ML investment decisions, pricing strategies, and policy.

Research into travelers' use of MLs to date has focused on the individual's willingness to pay a toll to realize travel time savings. ML choice models focus on the tradeoff between price (the toll) and value (usually, reduced travel time). More recently, researchers often include travel time reliability, arguing that all things being equal, travelers will pay for a more consistent travel experience. These hypotheses were confirmed by surveys of travelers who would, on paper, pay reasonable toll amounts to save time and improve their travel time reliability.

This basic model is built on logical, sensible assumptions and is workable in a limited data environment in which planners need few inputs beyond demographics, speed measurements, and traffic counts. However, there

[1]Texas A&M University, College Station, TX
[2]Cintra US, Austin, TX

**Corresponding Author:**
Address correspondence to John F. Brady: jbrady@cintra.us

now exist a few ML freeways for which trip data is available from both the GPLs and MLs, capturing individual traveler choices over a long time horizon. Surprisingly, these revealed choices do not confirm earlier hypotheses regarding lane choice. This research uses empirical trip data from both Katy Freeway MLs and NTE lanes to investigate lane choice on freeways with MLs. Novel findings include:

- Most travelers were not really choosing between lanes—they always used the same lane regardless of travel time and toll.
- Willingness to pay for travel time savings (expressed as toll paid divided by time saved on the MLs) was consistent across traveler groups when broken down by how frequently they traveled on the MLs.
- Travelers that did use both sets of lanes (termed "choosers") often made choices that appeared counter-intuitive based on travel time savings and toll rate.

This paper first reviews the state of the practice around ML use models, with a particular emphasis on how traveler lane choice is considered. The dataset and analysis is then introduced; a line of inquiry is followed and results noted. The paper concludes with highlights of this research and the authors' thoughts on how to proceed.

## Literature Review

Predicting traveler behavior is the cornerstone of good transportation planning. The ability to predict future travel patterns allows planners and engineers to better estimate the impacts of multiple transportation infrastrucutre investment options. This provides decision makers critical information needed to help them choose how to spend scarce investment dollars (*2*) and weigh policy options.

Travel time savings and the value of that time savings (VOT) to travelers are a critical benefit from infrastructure investment (*3*). How much travelers value their time spent in travel is also a dominant consideration in the prediction of future travel patterns. To demonstrate, most travel demand models determine route choice by putting drivers on the path which minimizes the individual's travel cost and that travel cost is dominated by the cost of travel time. Therefore, estimating and understanding how drivers value, compare, and select from among travel options (or choose not to travel) is the focus of a tremendous amount of research. In theory, data on travelers choosing MLs or GPLs would add to this knowledge–but our data raised more questions than it did to help derive a VOT.

*Value of Time.* The first analysis of VOT found was a 1925 Bureau of Public Roads report (*4*), which estimated VOT to be \$3/h in 1925 dollars. If adjusted for inflation to 2017 dollars, that would be a VOT of approximately \$41/h. Most early estimates were calculated in one of two methods:

1. Assume that VOT is proportional to the traveler's wage rate.
2. Estimate how much travelers would be willing to pay to use a faster mode of travel. For example, a traveler might be asked, "How much more would you pay to travel by car than by bus?"

VOT studies have found a strong relationship between the traveler's hourly wage rate and his or her VOT (*5*). Various studies have estimated VOTs from 20% to 100% of the traveler's hourly wage rate. However, most literature has suggested that the VOT should be around 50% of the hourly wage rate for personal trips (*6*). For commercial trips, VOT may be as high as 1.7 times the average wage rate (*7*).

The majority of research efforts in the last 40 years have sought to update VOT using stated-preference (SP) surveys. Travelers being surveyed are generally presented with a set of predetermined, hypothetical travel alternatives and asked to state their preference. For example, would you prefer Option 1, which takes 10 min and requires a US\$2 toll, or Option 2, which takes 15 min but has no toll? (*8*). The results of these surveys are used to develop logit equations that predict mode choice and estimate VOT. Carrion and Levinson (*8*) and Concas and Kolpakov (*5*) provide good overviews of VOT research.

To emphasize the extent to which SP VOT studies are relied upon in the practice of travel prediction, the U.S. Department of Transportation recommends values of time of US\$10.60/h for commuter travel and US\$21.46/h for business travel. The latest update to these guidelines (*9*) suggested a VOT of US\$12.50 in 2009 dollars for all purposes of travel combined. The United States is not alone as many countries have incorporated VOT into their economic evaluation of transportation projects.

There are far fewer studies that have used revealed preference (RP) data, where travelers' real-world choices are relied upon for all or part of the data used in the VOT analysis. Usually, SP and RP findings tend to differ, with RP studies often finding higher VOT. Researchers in California have examined travelers on SR-91 Express Lanes and the I-15 HOT lanes to determine VOT and value of travel time reliability based on real-world travel on those lanes. They combine RP data with some SP (survey) data in their analyses. Brownstone and Small use the combined data to find VOTs in the range of US\$20/h (*10*); they, and Ghosh (*11*), note that the SP studies find VOT to be much lower than in RP studies. For example, Devarasetty et al. found a fairly high (US$50/h) VOT for Katy Freeway ML travelers using combined SP and RP data (*12*). One popular hypothesis for the differences between SP and RP VOT is that travelers misperceive their travel times and costs. Devarasetty et al. examined Katy Freeway travelers using both RP and SP data; they found that responses to the question of how much time they saved by using the MLs overestimated actual travel time savings by a factor of three (*13*).

Despite their differences, both SP and RP studies to date have examined driver choices as discrete events, decisions disconnected from the driver's medium-term habits and experiences. Regional transportation models typically model a single day, deriving seasonal and annual results by applying scaling and annualization factors.

### *Managed Lanes and VOT*

An operational ML facility provides the perfect opportunity to study driver responses to travel choices: pay to travel on the (generally) faster MLs or travel on the (generally) slower GPLs toll-free. Not surprisingly, considerable research has been undertaken to quantify traveler's VOT on ML facilities, utilizing both SP- and RP-based approaches. This includes several papers by one of the authors who surveyed ML travelers to better understand their VOT (*12–14*). These VOTs were generally reasonable, although sometimes higher than in the existing literature.

VOT analysis is of particular interest in ML investment decision-making. Many traffic and revenue forecasting firms, along with toll-road operators, spend considerable time and resources estimating the potential use of MLs based on VOT. Before construction begins, the modeled revenue stream is instrumental in securing financing for the project. Once operational, these models may be relied upon to inform pricing strategy or design changes.

Some meta-analyses of toll-road forecasts have found that traffic and revenues are often overestimated (*15*). However, estimates are tending to become better and most are within the range of actual revenues so that the facility can meet its financial obligations. To our knowledge, all of the recent major Public-Private Partnership ML projects that are in operation today have relied heavily on traffic and revenue forecasts modeled using VOT, including NTE (*16*), the I-495 Capital Beltway (*17*) and numerous others that have been more recently bid. At the time of writing, each of these projects have seen their debt ratings maintained or raised by rating agencies (*18*, *19*), suggesting that on the whole, this aggregate approach has succeeded in approximating the revenue-generating potential of MLs. Recent theoretical work has begun to take into account some driver behavior that is rational but might not follow traditional model assumptions. For example, Lou et al. examined ML trips using boundedly rational user equilibrium in which travelers may not adjust their trip for small travel time savings (*20*). This is a logical extension of previous modeling work and brings an important human element to the model.

Therefore, current methods appear to produce reasonable estimates of traffic and revenue in aggregate, and therefore one would assume have estimated a reasonable VOT. However, the data examined in our study finds the standard VOT-based assumptions, and even the advanced analysis concepts, do not hold when examining millions of individual driver choices.

With most of the VOT results derived from SP studies, there is some skepticism on how well those survey responses translate into real-world travel choices. However, until now no one was able to look at individual travelers over time and examine how each was choosing between the paid MLs and the toll-free GPLs over time. With the rather unique datasets from Katy Freeway MLs and NTE (discussed in the next section) this research examined millions of travelers' use of the MLs and how much they were willing to pay for the use of the lanes and the time they saved. In theory, this would shed additional light on travelers' VOT. However, these results were particularly surprising and contradict much of what was expected based on past research.

## Data

Data were obtained from the Katy Freeway and NTE in Texas. For both facilities, the MLs operated at free-flow or near free-flow speeds with rare exception and the GPLs would routinely become congested during peak hours.

The following sections describe the facilities and discuss the methods by which each dataset was collected and synthesized. They are quite similar, with a couple key distinctions between the projects:

- Display of travel times: Dynamic message signs display the GPL and ML travel times for key destinations along the Katy Freeway corridor. At NTE, travel times were not displayed during the study period. On both facilities, the ML conditions are easily visible from the GPL.
- Speed limit: Both MLs and GPLs have the same speed limit on Katy Freeway. At NTE, the MLs have a posted speed limit 10 mph higher than the GPLs. This difference gives the NTE a travel time advantage when both sets of lanes are uncongested.

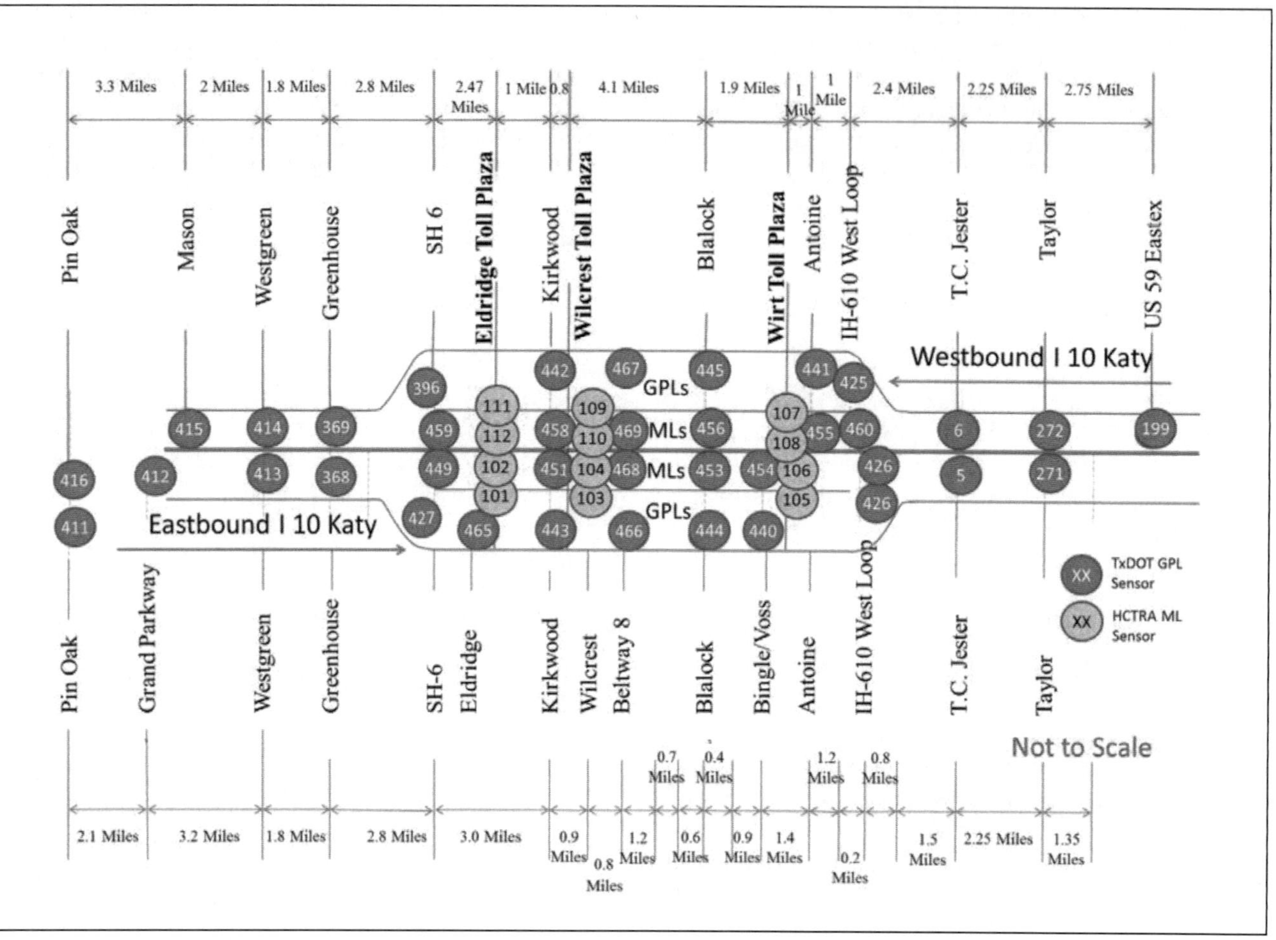


**Figure 1.** AVI sensor locations along Katy Freeway.

## *Katy Freeway Data*

The I-10 Katy Freeway connects the city of Katy to downtown Houston. This 12-mi section of freeway has up to six GPLs and two variably priced MLs in each direction. The Katy Freeway MLs generally require less travel time and are usually more reliable than the adjacent GPLs. The MLs are separated from the GPLs by flexible pylons. Drivers are required to pay a toll that varies depending on the time of day and the number of people in the vehicle. HOVs with two or more occupants and motorcycles can use MLs free of charge during HOV-free hours. HOV-free hours are Monday through Friday from 5 a.m. to 11 a.m. and 2 p.m. to 8 p.m. HOVs and motorcycles pay the same toll as single-occupant vehicles at all other times. To avoid the toll during the HOV-free hours, HOVs and motorcycles must pass the toll plazas in the HOV lane, the leftmost lane of MLs.

Texas Department of Transportation (TxDOT) operates automatic vehicle identification (AVI) sensors located on both the MLs and GPLs along Katy Freeway. Figure 1 shows the location of the sensors, with each number indicating a specific sensor. These sensors detect vehicles with transponders (or tags) and records the unique transponder ID from the vehicle and time of detection. All vehicles that pay a toll on the MLs are required to use a transponder. Many other vehicles traveling on Katy Freeway also have transponders. The AVI data obtained from TxDOT consists of all sensor detection records from most of 2012, 2013, and 2014. The bulk of the analysis focuses on March, April, and May of 2014 so that it was more readily comparable with the 3 months of NTE data available.

The Harris County Toll Road Authority (HCTRA) operates the MLs and collects AVI data along the MLs at the three toll plaza locations (see Figure 1 for HCTRA sensors). These data obtained from HCTRA included the unique transponder ID, date and time of record, and toll paid (if applicable) for each vehicle traveling within the MLs for 2012, 2013, and 2014 (only through September because the data were obtained in October 2014). Travelers who did not pay to use the MLs (including HOVs during rush hours and exempt public vehicles) were not included in the analysis.

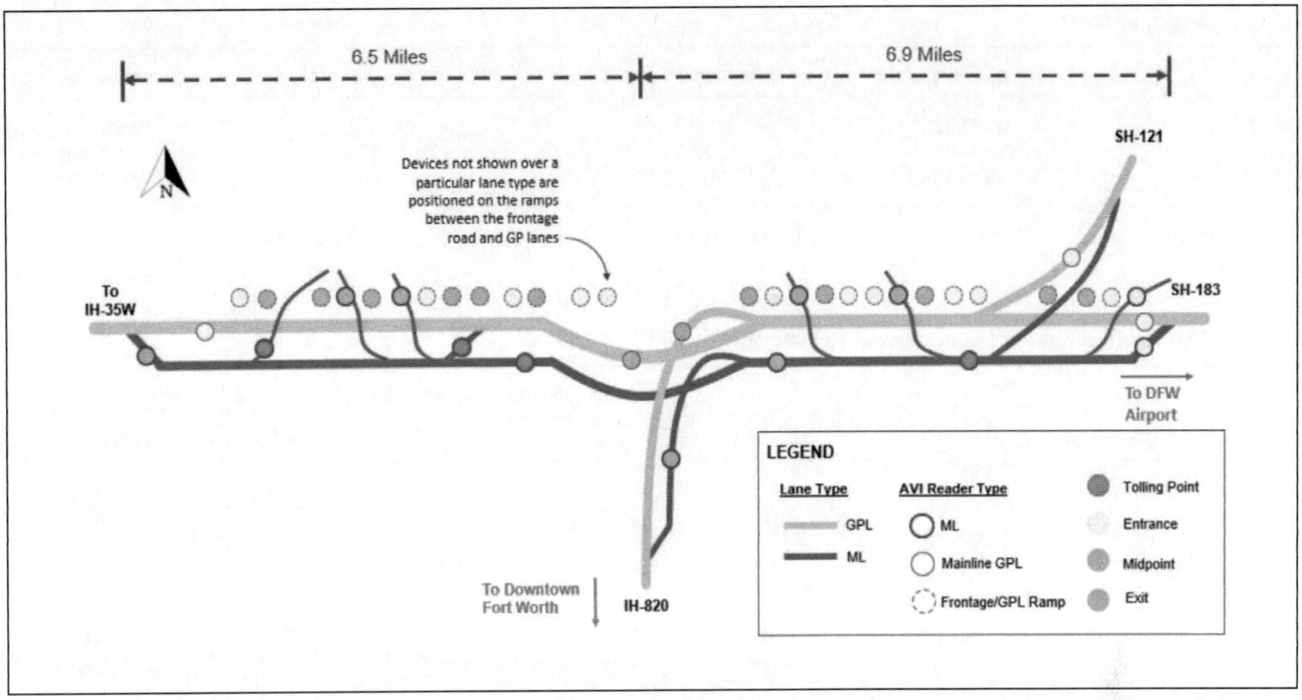


**Figure 2.** AVI sensor locations along North Tarrant Express, westbound shown.

To ensure that no transponder owner could be identified using the transponder IDs, each transponder ID was assigned a unique random ID, and the original transponder IDs were deleted. Therefore, the dataset could never be used to identify specific individuals traveling on Katy Freeway. The dataset could still be used to track the trips of vehicles throughout the 3 years based on the random ID that each vehicle was assigned.

Travel time and distance traveled by freeway travelers were calculated using the time and location of the sequential detection of unique IDs. Based on the time of detection and the toll schedule, tolls were assigned to the trips that were detected at toll plazas in the MLs.

To estimate how much travelers value their travel time, a simulated trip on the alternative lane was created for each observed trip. For observed trips on the toll lane, the simulated trip was free on the GPLs, and vice-versa. Simulated trips started and ended at the same point on the freeway as the real trip, just in the opposite lane type. The simulated trip travel times were constructed from observed trips made by other vehicles within the same 15-min interval on the same day the observed trip was made. For a small percentage of trips (less than 2% in the peak and shoulder, and 8% in the off-peak), no observed trips were available for the alternate lanes so a historical average was used.

By far the largest source of unusable data came from drivers that alternated between the ML and GPL during a single trip, as it was impossible to determine their travel time savings. This occurred in approximately 15% of trips and these trips were not included in the analysis. Quality control on the data did reveal other minor problems and some trips were unusable. Creation of these trips was a challenging process and more details are available in Sunghoon Lee's dissertation (*21*).

The final dataset contained two records for each trip. The two records represented the two choices for the trip: the one that was made and the one on the lanes not chosen. The trip parameters included in the final dataset are the random ID, lane choice, travel time, total toll paid, and several variables not used in this research.

## NTE Data

The NTE corridor is an east–west oriented 13-mi stretch of highway that serves as a circumferential road around Fort Worth, TX and a primary highway route to access the Dallas/Fort Worth International Airport. Redevelopment of the corridor began in 2011, with works overseen by NTE Mobility Partners, LLC, a private consortium led by Cintra under a comprehensive development agreement with TxDOT. Since construction was completed in 2014, each direction contains two to three GPLs, two tolled MLs, and continuous frontage roads. The MLs are separated from the GPLs by a concrete barrier, effectively operating as a wholly separate highway-within-a-highway.

Vehicles that choose the MLs are required to pay a toll that varies dynamically in real time. Non-tag-equipped

**Table 1.** Size of Datasets

| Freeway | Number of transponders | Number of GPL trips | Number of ML trips | Total number of trips |
|---|---|---|---|---|
| Katy Freeway | 1,108,002 | 9,197,423 | 800,878 | 9,998,301 |
| NTE | 884,443 | 11,414,972 | 2,624,126 | 14,039,098 |
| TOTAL | 1,992,445 | 20,612,395 | 3,425,004 | 24,037,399 |

*Note*: Katy Freeway = March, April, and May of 2014. North Tarrant Express (NTE) = August, September, and October of 2015.

**Table 2.** Percentage of Travelers Choosing MLs by Frequency of Travel

| | | Share of trips on the MLs | | | |
|---|---|---|---|---|---|
| Frequency of freeway travel | Roadway | 0% | 1% to 99% | 100% | Number of tags |
| Infrequent (1–3 trips per month) | Katy | 89.3% | 6.8% | 3.9% | 894,408 |
| | NTE | 52.9% | 26.9% | 20.2% | 644,106 |
| Sometimes (4–12 per month) | Katy | 68.5% | 30.6% | 0.9% | 152,390 |
| | NTE | 35.1% | 61.9% | 3.0% | 137,566 |
| Commuter level 1 (13–29 per month) | Katy | 51.9% | 47.1% | 1.0% | 51,684 |
| | NTE | 28.8% | 69.8% | 1.4% | 62,701 |
| Commuter level 2 (30 or more per month) | Katy | 55.8% | 43.5% | 0.7% | 9,520 |
| | NTE | 33.3% | 65.9% | 0.8% | 40,070 |
| TOTAL | Katy | 84.4% | 12.3% | 3.3% | 1,108,002 |
| | NTE | 47.5% | 37.2% | 15.3% | 884,443 |

*Note*: Katy Freeway = March, April, and May of 2014. North Tarrant Express (NTE) = August, September, and October of 2015.

vehicles and trucks may use the MLs for a higher toll rate. HOVs with two or more occupants and motorcycles can use the MLs for a 50% discount during rush hour and must pay the same toll as single-occupant vehicles at all other times. To receive the discount, vehicles must electronically self-declare before their trip using a mobile app.

NTE maintains AVI sensors located on both the MLs and GPLs along the corridor. Figure 2 depicts the arrangement of the sensors, with each circle indicating a specific sensor array that includes at least one AVI positioned over each lane of travel. The data obtained from the AVI readers for this study consists of all sensor detection records from August to October 2015, a 3-month time period with high-quality data.

To ensure privacy, the transponder ID records were stored on a secure server and the data anonymized and aggregated before being analyzed. The customer-specific information associated with the tag was maintained by the tag issuer to preclude the possibility of personally identifying customers.

At the NTE, detections within a 30-min window for a unique tag were matched to create a trip. If both the origin and destination were detected, as was usually the case, the lane choice was recorded and the travel time calculated. If the trip could have been made using the lanes that were not chosen by the traveler (the alternate lanes), the travel time for the alternative was calculated by taking the average travel time among drivers who chose the alternative lane within a $+/-$ 5-min window. Trips that utilized the MLs had the toll amount assigned to the trip using the toll(s) charged to the vehicle. If a tag was only detected at one location, the "trip" was recorded, along with the lane choice, and toll (where applicable).

In all, the final dataset contained one record for each trip. The trip parameters considered in this study were the vehicle's unique ID, trip start time, lane choice, travel time on the chosen path, toll paid on the chosen path, travel time on the alternative path, and toll on the alternative path. Note that non-tag-equipped vehicles, as they are not detected by AVIs, were not included in this dataset. Tag IDs identified as trucks were removed. HOV self-declared trips account for about 2% of rush-hour transactions at the NTE facility.

Both freeways were examined over a 3-month time period. For Katy Freeway this was March, April, and May of 2014. These were the last three continuous months of data available for this freeway that did not include the lower volume summer months. For NTE this

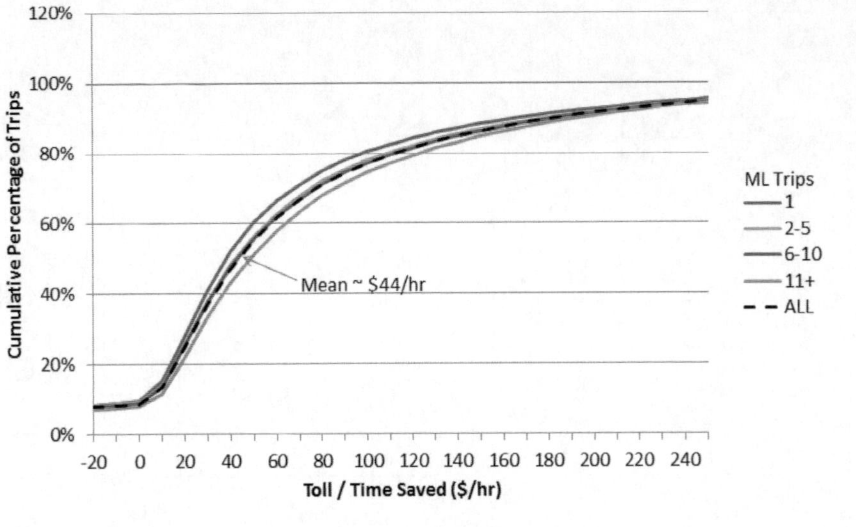


**Figure 3.** Willingness to pay to use Katy Freeway MLs by frequency of ML travel.

included August, September, and October of 2015. This period of time was chosen for the high quality of data available in that window.

## Results

The results of this examination of real-world travel choices on MLs began with an examination of the size of the datasets used in the analysis (see Table 1). Previous SP and RP studies generally survey a few hundred to a few thousand travelers at most. Prior RP studies were conducted on datasets approximately that size (*10–12*). This dataset includes nearly 2 million unique travelers (transponders) making over 24 million trips across a 3-month period. Therefore, the size of the dataset is orders of magnitude larger than preceding studies and includes the vast majority of vehicles who paid to use the MLs. It uniquely considers data from two different roadways in two different cities. If the observed travel behavior is similar on both facilities then observed deviations from long-held assumptions must not be, at the very least, attributable to the unique design, driver sentiment, small sample size, or the specifics of a single urban corridor. Table 1 describes the size of the dataset.

In theory, all tag-equipped vehicles on the Katy Freeway and NTE should consider which set of lanes to choose for each trip. The existing literature considers travel time saved, travel time reliability, and price among the primary exogenous factors motivating choice. The traveler's perceived value of these factors, combined with their willingness to pay and individual trip purpose should largely explain their choice. Therefore, it would be reasonable to assume that a frequent user making dozens of trips across a 3-month time frame would experience a combination of travel time savings, reliability, toll, and trip characteristics (such as trip purpose) that would result in the choice to utilize the MLs at least once. However, this is certainly not the case, especially on Katy Freeway.

Table 2 shows how frequently a traveler (a unique tag ID) is observed to use the MLs given how frequently they travel on the freeway (ML and GPL combined). Using the Katy Freeway as an example, the most frequent travelers made over 30 trips per month (90 or more total trips) on the freeway. Approximately 56% of those frequent travelers did not use the MLs at all, 43% used both MLs and GPLs, whereas just under 1% used the MLs exclusively. Therefore, almost half of this group is clearly considering MLs—but this group represents only 4.4% of the transponders observed on the freeway. Most travelers used the Katy Freeway one to three times per month (infrequent users) and only 7% of those travelers used both sets of lanes during the 3-month period. For all Katy Freeway travelers combined, only 12% were "choosers"—those that selected both lane types at least once. At NTE there were relatively more choosers across all frequency groups, however, a sizable number did not use the MLs at all during the study period—we observed that 48% of all travelers and 33% of the most frequent travelers at NTE exclusively used the GPL.

The frequency with which the travelers used the freeway did not appear to substantially influence their likelihood of being choosers. With the exception of the infrequent group, travelers tended to exhibit a similar likelihood of being choosers at each facility. These data are only transponder reads and there is no information

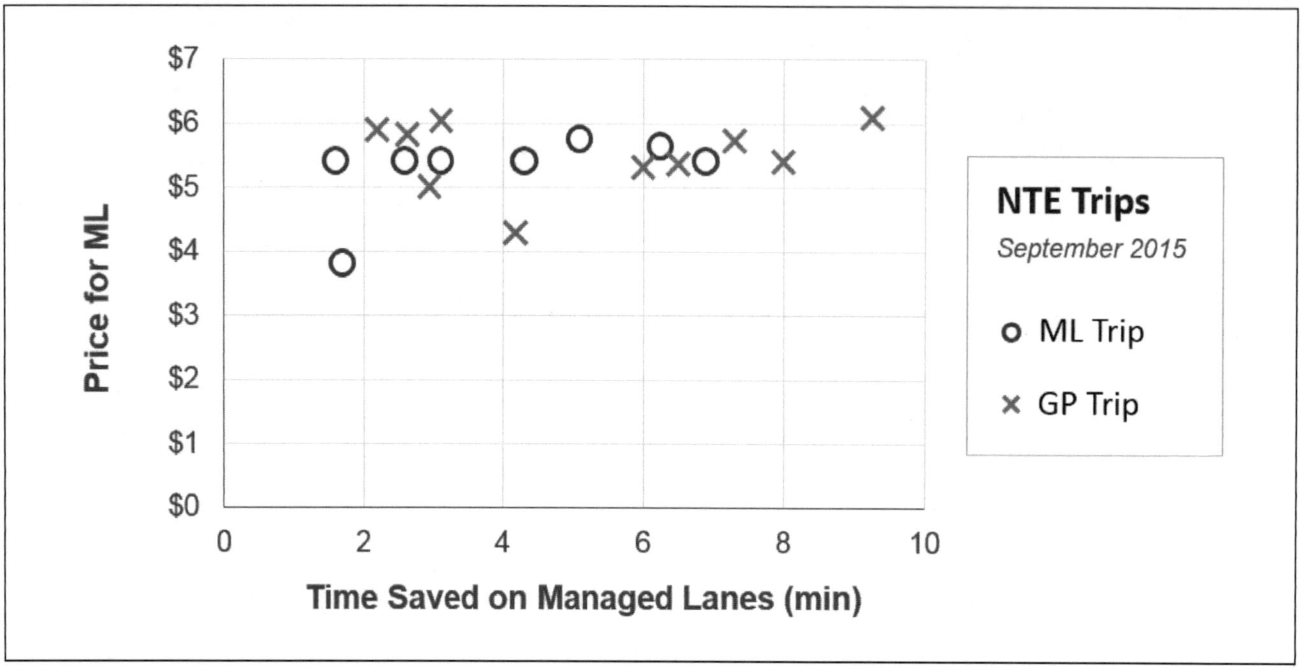


**Figure 4.** Lane choices made during the PM rush hour by a single NTE driver in September 2015.

on the users, their trip purposes, trip urgency, or even if it was the same driver (for example, it could be different family members driving on different trips) each time the transponder was identified.

Further insights into the reasons behind a traveler's choice cannot be derived from this data and so we posit two hypotheses. The first relates to decision aversion. Some travelers do not want to be faced with making a decision on each trip. Possibly even worse would be the feeling you have made a bad decision—taking the MLs when the GPLs were about the same speed or taking the GPLs when they were heavily congested. A traveler who has decided to always take the GPLs (for example) is no longer faced with a decision and can therefore no longer make a bad decision. A second hypothesis is that infrequent users may not fully understand how to use the lanes, where to enter/exit, or when to get the best value (receive substantial time savings for the toll paid). By contrast, perhaps the more frequent users of the freeway are more likely to be choosers because they have a better understanding of these same concepts. This hypothesis can be examined by looking at the toll paid versus the travel time saved for these individual groups of travelers (see Figure 3).

Figure 3 displays the toll amount paid for each trip on the MLs divided by the travel time saved by using the MLs. We use this as a surrogate measure for a traveler's willingness to pay as it shows that the traveler opted to pay at least that much. Additionally, it only represents those travelers who did choose to pay the toll. Other travelers might have been willing to pay a toll, just not quite as much as the toll that was charged at the time of their travel. Finally, it implies that the traveler would know how much travel time they would save. It is likely that the traveler would have a rough idea of their travel time savings (see data section), but probably not the exact amount.

Nevertheless, this surrogate willingness to pay was remarkably consistent across traveler groups. Frequent travelers (termed "commuters") ended up getting the same value for their toll dollar as "sometimes" travelers and "infrequent" travelers. This would suggest either that frequency of travel did not improve the traveler's ability to use the MLs to maximum advantage (largest time savings for smallest toll) or that the travelers are not trying to maximize this discrete value proposition on each trip. Anecdotal evidence suggests such an analysis is a far greater cognitive load than most travelers are willing to spend on planning their trip.

Furthermore, approximately 9% of ML trips across all groups occurred when the MLs took longer and this resulted in a negative willingness to pay. Clearly the traveler did not want to pay to go slower, but chose the MLs for other reasons. When looking at the data in aggregate as in Figure 3 it is difficult to delve any further into this line of inquiry. Therefore, the toll paid and travel time saved was next examined at a disaggregate level.

Figure 4 below summarizes a randomly selected commuter's workday PM peak-period trip choices observed across a 1-month period in September 2015 on the NTE. Each mark represents a trip with the choice of lane indicated by the shape.

What is remarkable about this pattern, which we also observed among choosers in aggregate, is that the price and time savings, even across a wide range, offers little help in predicting the individual traveler's lane preference for a specific trip. This particular driver was presented with prices between US$3.75 and US$6.10 and time savings between a few minutes and about 10 min. If the prevailing normative theory held at the individual level, this driver would have a higher affinity for the MLs for trips on the right-hand side of the figure (similar price, more time saved).

Without additional data we, again, can only posit hypotheses about this observation. One is that the driver's revealed preference for the MLs—in this specific case, 8 out of 18 trips—is not the result of 18 discrete choices, carefully informed by trip purpose, price, value of reliability, and travel time, but rather an unobservable change that outweighs the classic VOT assumptions. For example, sometimes there is an event (for example, their turn to cook dinner) so the user must be on time whereas other days there is flexibility in arrival time. Or, possibly, it could simply be the mood of the traveler. Some days they do not feel like dealing with GPL traffic or they feel like giving themselves the luxury of a free-flow trip home (we assume home, based on the regular PM occurrence of this trip).

## Conclusions

> I call it theory-induced blindness: once you have accepted a theory and used it as a tool in your thinking, it is extraordinarily difficult to notice its flaws. If you come upon an observation that does not seem to fit the model, you assume that there must be a perfectly good explanation that you are somehow missing. ... Disbelieving is hard work.
>
> Daniel Kahneman. *Thinking, Fast and Slow*

A favorite interview question of one of the authors is "Imagine that a person used the NTE highway yesterday and I want your help guessing if they chose the ML or GPL. You can ask me (the interviewer) or the driver for any information to help you guess. What would you ask?" The reader may not be surprised to learn that the vast majority of students of transportation, without blinking, ask about the level of congestion yesterday and ask the driver how much money they make, if they had a passenger in the car, the purpose of their trip, and how many children they have. These are among the typical inputs to a model of ML use. It is a case of theory-induced blindness. Compared with the depth of information within the student's grasp, his proposed solution is shallow. He sticks with what he knows and passes up the opportunity to ask the questions that students of other disciplines tend to come up with first, like "Do you often use the express lanes?"

After a preliminary review of this dataset, we have arrived at a few conclusions. First, many travelers on priced ML corridors are not choosing between the lanes. This finding on its own is not inconsistent with the classic VOT model or even more modern bounded rational behavior models, but it is notable that a substantial number of apparently informed, frequent drivers are sticking to one set of lanes across diverse circumstances. Several hypotheses were posited, including lack of understanding of the MLs and decision aversion. A better understanding of how to predict whether a traveler is a "chooser" or "non-chooser" would be a significant step forward in predicting the use of MLs.

Second, predicting when a chooser will select MLs versus GPLs is not straightforward. We set aside the question of the price-time savings tradeoff and observe that frequent drivers of the highway do not exhibit behaviors that suggests they are "optimizing" their lane choice any more than the very rare user. Third and finally, individual lane preferences observed over time suggest choosers do not seek to "optimize" each trip, but rather select a mix of lane choices that does not correspond to travel time savings and toll paid. How such decisions are made should be an area of intense future research and would be critical in developing an alternative, cohort-centric approach to understanding and forecasting ML use over time.

### Acknowledgments

The authors thank Texas Department of Transportation, Houston TranStar, and the Harris County Toll Road Authority for providing the data for Katy Freeway. Many Texas A&M University students helped with Katy Freeway data cleaning and database development, including Sunghoon Lee and AKM Abir. These earlier efforts that included data cleaning and database development were funded in part by the Policy Research Center at Texas A&M Transportation Institute, the Texas Department of Transportation, and the Federal Highway Administration through the University Transportation Centers program. The funding from these sources is gratefully acknowledged. The authors thank the NTE technical team, especially Wei He, whose analytical techniques underpinned the NTE's analysis; and Ricardo Sanchez of Cintra for his foresight to install AVI devices at NTE.

### Author Contributions

The authors confirm contribution to the paper as follows. Study conception and design: MB, JB; data collection: MB, JB; analysis and interpretation of results: MB, JB; draft manuscript preparation: MB, JB. All authors reviewed the results and approved the final version of the manuscript.

## References


1. Florida Department of Transportation. *"95 Express Annual Report" Covering July 1, 2009 to June 30, 2010 and Covering July 1, 2011 Through June 30*, 2012, 2011, 2013, p. 7.
2. Meyer, M., and E. Miller. *Urban Transportation Planning*, 2nd ed. McGraw Hill Higher Education, New York City, NY, 2001.
3. United States Department of Transportation. *Revised Departmental Guidance on Valuation of Travel Time in Economic Analysis, by Peter Belenky*, 2011. https://www.transportation.gov/sites/dot.dev/files/docs/vot_guidance_092811c.pdf. Accessed July 20, 2017.
4. U.S. Bureau of Public Roads and the Cook County Highway Department. *A Study of Highway Traffic and the Highways System of Cook County*. U.S. Bureau of Public Roads and the Cook County Highway Department, Orland Park, IL, 1925.
5. Concas, S., and A. Kolpakov. *Synthesis of Research on Value of Time and Value of Reliability*. No. BD 549-37, 2009.
6. Small, K. *Urban Transportation Economics*. Harwood Academic Publishers, Chur, Switzerland, 1992.
7. Waters, W. *Value of Time Savings for the Economic Evaluation of Highway Investments in British Columbia*. Reports and Studies, British Columbia Ministry of Transportation, Victoria, Canada, 1992.
8. Carrion, C., and D. Levinson. Value of Travel Time Reliability: A Review of Current Evidence. *Transportation Research Part A: Policy and Practice*, Vol. 48, 2012, pp. 720–741.
9. Ayala, R. *Revised Departmental Guidance on Valuation of Travel Time in Economic Analysis*. Technical report. U.S. Department of Transportation, Washington, D.C., 2014. http://www.transportation.gov/sites/dot.gov/files/docs/USDOT%20VOT%20Guidance%202014.pdf.
10. Brownstone, D., and K. A. Small. Valuing Time and Reliability: Assessing the Evidence from Road Pricing Demonstrations. *Transportation Research Part A: Policy and Practice*, Vol. 39, 2005, pp. 279–293.
11. Ghosh, A. *Valuing Time and Reliability: Commuters Mode Choice from a Real Time Congestion Pricing Experiment*. PhD dissertation. Department of Economics, University of California at Irvine, Irvine, CA, 2001.
12. Devarasetty, P., M. Burris, and W. Shaw. The Value of Travel Time and Reliability: Evidence from a Stated Preference Survey and Actual Usage. *Journal Transportation Research Part A: Policy and Practice*, Vol. 46, No. 8, 2012, pp. 1227–1240.
13. Devarasetty, P., M. Burris, and C. Huang. Comparing Perceived and Actual Travel Time Savings on Freeways with Managed Lanes. *The Journal of Transportation of the Institute of Transportation Engineers*, Vol. 6, No. 1, 2014, pp. 1–13.
14. Patil, S., M. Burris, D. Shaw, and S. Concas. Variation in the Value of Travel Time Savings and Its Impact on the Benefits of Managed Lanes. *Transportation Planning and Technology*, Vol. 34, No. 6, 2011, pp. 547–567.
15. Bain, R. Error and Optimism Bias in Toll Road Traffic Forecasts. *Transportation*, Vol. 36, No. 5, 2009, pp. 469–482.
16. NTE Mobility Partners. *LLC Texas Private Activity Bonds Official Statement*, p. 535. https://emma.msrb.org/EP367750-EP288778-EP684048.pdf. Accessed July 28, 2017.
17. Capital Beltway Funding Corporation of Virginia. *Toll Revenue Bonds Official Statement*, p. 454. https://emma.msrb.org/MS65630-MS269419-MD533919.pdf. Accessed July 28, 2017.
18. Moody's Affirms NTE Mobility Partners LLC's Baa2 Senior PAB. https://www.moodys.com/research/Moodys-affirms-NTE-Mobility-Partners-LLCs-Baa2-senior-PAB-and-PR_350038?WT.mc_id=AM˜RmluYW56ZW4ubmV0X1JTQl9SYXRpbmdzX05ld3NfTm9fVHJhbnNsYXRpb25z~20160711_PR_350038. Accessed July 28, 2017.
19. Transurban Primer. https://www.transurban.com/content/dam/investor-centre/toolkit/transurban-overview.pdf. Accessed July 28, 2017, p. 3.
20. Lou, Y., Y. Yin, and S. Lawphongpanich. Robust Congestion Pricing Under Boundedly Rational User Equilibrium. *Transportation Research Part B: Methodological*, Vol. 44, No. 1, 2010, pp. 15–28.
21. Lee, S. *Real Option Analysis to Value Managed Lanes Using Big Data*. Dissertation. Department of Civil Engineering, Texas A&M University, College Station, TX, 2016.


*The Standing Committee on Congestion Pricing (ABE25) peer-reviewed this paper (18-04294).*

*Any errors or omissions are the fault of the authors.*